SMART LAND SCAPE

GIULIA GARBARINI

INDEX

FOREWORD by Mosè **Ricci**		004
START WITH SMART		008
ABSTRACT		008
INTRODUCTION		010

1 FIVE CONCEPTS OF LANDSCAPE — 016

1.1	LANDSCAPE 0.0 NATURE	023
1.2	LANDSCAPE 1.0 REFUGE	028
1.3	LANDSCAPE 2.0 FACTORY	033
1.4	LANDSCAPE 3.0 SQUARE	038
1.5	LANDSCAPE 4.0 NETWORK	043

2 KEY WORDS AND TIMELINE — 052

2.1	KEY WORDS	054
2.2	TIME-TABLE	058

3 SMART LANDSCAPE — 063

3.1	SMART	065
3.2	CITY	070
3.3	LAND	074
3.4	NETWORK	078
3.5	LANDSCAPE	084
3.5.1	LEARNING FROM THE STRATEGY OF SURFACE	088
3.5.2	DRIVERS TOWARDS A SMART AND ADAPTIVE LANDSCAPE	093

4 REFERENCE — 100

4.1	**TOOLS AND DEVICES**	104
4.2	**REFERENCE NETWORK**	109
4.2.1	**SMART GRID**	125
4.3	**REFERENCE PROJECT**	116
4.3.1	**REFERENCE - BACKGROUND PROJECT**	117
4.3.2	**REFERENCE - MAIN PROJECT**	132
4.3.2.1	**TÅSINGE SQUARE**	134
4.3.2.2	**SOUTH-EST COSTAL PARK PHOTOVOLTAIC PERGOLA**	146
4.3.2.3	**VÈ-LIB AND AUTO-LIB**	152
4.3.2.4	**ECOGRID BORNHOLM**	166

5 TRANSFER OF KNOWLEDGE — 178

6 BIBLIOGRAPHY — 186

LANDSCAPES 4.0

FOREWORD by Mosè Ricci

"... It is the crisis that forces us to invent new forms of interpreting reality. And landscape, at least until now, is the only possible one. It is the form of an area and it has taken place of other concepts, such as that of space, putting perception first. ..."

Franco Farinelli

Simultaneous action of three decisive factors – economic and environmental crises and revolution of information sharing technologies – is changing manner both our lifestyles and the way we imagine and desire the solid forms of our future in such a profound that all our projecting knowledge suddenly becomes inadequate both as an interpretative instrument of the present condition and as a device capable of generating new aesthetic qualities, environmental, social, and economic performances. In fact, great technological changes have always produced great transformations in the ways and forms of living and, consequently, in the ways and forms of projecting. One of the main theoretical issues of modernity was that of the best possible spatial synthesis between function and form. Today, with the revolution of information technologies, we are facing the opposite problem: to give meaning, narrative and uses – even temporary – to spaces that already have got some given forms, and to turn them into places where living is both pleasant and affordable.

This phase requires of designers to adapt new points of view on the future (paradigms, as Thomas Kuhn would say), but also and above all on the present, and a new idea of projecting a physical space. This is a challenge that puts in value what already exists. A challenge that considers a context as a project, a landscape as a new infrastructure, which produces values for living, and the present of a city as a collective and non-authorial picture.

Landscape is the mother of new project paradigms in the age of information sharing. In a way, it contains them all. With great efficacy, Franco Farinelli makes it clear in his article "Return of the landscape" in a Corriere della Sera (Sunday, December 20, 2015), where he states that "landscape today is the only possible cognitive model (...) exactly the same way as in front of a landscape, within a network distance between a subject and an object is no longer possible, space is no longer thinkable (...) In short: because it is the opposite of space. The cognitive model of landscape (based on totality and mobility of a subject as well as on its indissolubleness towards an object) is the only one that brings together epistemological conditions imposed today by our world, and that on our side is urgent to recognize and assume if we want to try to understand how the world functions."

Landscape is the only material and conceptual context of a physical space project. Landscape replaces the modern idea of territory as a projection of the reality on a horizontal plane of representation with a single linear metric measure, because distance is no longer possible within landscape as well as within a network. Landscape, like a network, is round. Like a network, it is defined through the multiplicity of gazes and actions. Charles Waldheim writes in Landscape as Urbanism (Princeton University Press, 2016) that it is possible to compare modern cities to the Piranesi's *disabitato*. A seemingly informal place where nature and traces of previous eras compose a landscape full of meanings and people. Places, as well as tools and project devices, cannot remain the same in the city beyond metropolis. And the new plan of Detroit – of Stoss Landscape Urbanism – (Chris Reed, one of the partners of this research, was also in the group of Stalking Detroit) abandons the ideas of zoning and expansion in favour of goals calibrated on the ecological performances of new urban materials. It is in fact conceived as a map, an application for a smartphone.

Within a strategic framework that defines possible spatial structures of networks of productive landscapes in a city (food, energy, water, pollution abatement etc.) the plan presents alternative possibilities of specific interventions on landscape to its inhabitants. Everyone can choose to activate an option they consider more favourable and get environmental and social benefits that represent objectives of change on an urban level. The result of this process provides sense and beauty to a new form of city-landscape where buildings can become trees, nature is the main infrastructure that connects citizens to the quality of life, and landscape architecture is a sensitive form of living in the age of digital revolution after modernity.

Modernity has offered us a vision of landscape as a field of definition of a city's physical space or as an open space in a city, but, as Joao Nunes says, landscape is like a constantly moving portrait of a society that inhabits it. In this sense, failures were perhaps more useful than successes. The end of modernity seems to be leading us towards a holistic conception of landscape, which is becoming a device. It is like an application, a software that allows us to use qualities of a living space in a better way, and also, in its most updated version, to repair its defects. This seems to be the lesson we can learn in the cities of the Western world affected by the crisis and by their redemption trajectories that are finally starting to outline some important changes in the paradigms of urban quality.

Nowadays the question resides in the transition from a new way of thinking about landscape to a new way of designing Landscape. If, as Franco Farinelli suggests, we can conceive landscape as a physical context of contemporary living more similar to a digital network, perhaps we can just conceive it as a network. As if the utopian visions of Superstudio's Supersurface or Agronica by Andrea Branzi unexpectedly became real, physical facts.

It is not news. Energy Smart grids are already using geographic contexts as areas for development and measure the potential of settlement landscapes. But to think of landscape as a network means a possibility to move an infinite series of data along a grid of reference for the existing space and to be able to localize its nodes in specific physical places capable of catalyzing the existing physical space making it valuable and proactive, in short, valorizing it.

Within this thesis, we are working on different academic sites with applied studies, project experiments and theoretical contributions. The research "Smart landscape" conducted by Giulia Garbarini and described in this volume, in combination with the volume dedicated to a case study "L.I.D.O. – Learning Island Design Opportunities" developed on the island of Venice Lido, which accompanied the research, today represent the most advanced point of the undertaken scientific path.

1.

"In chimica i catalizzatori sono sostanze che, presenti pur in minima quantità, esercitano sulla velocità di una reazione un'azione accelerante (o ritardante se catalizzatori di segno negativo) prendendo parte agli stadî più importanti della reazione stessa, e poi rigenerandosi, per ritrovarsi così inalterati alla fine del processo.In questo caso il termine catalizzatore (Oswalt P 2013) è usato, nella sola accezione positiva, per identificare quei luoghi esistenti che, in seguito all'attivazione o alla realizzazione di alcuni DISPOSITIVI (tecnologici, gestionali, sociali, urbani etc.) possono favorire o accelerare la trasformazione." C. Rizzi, "Background – Purpose", in L.I.D.O. - LEARNING ISLAND DESIGN OPPORTUNITIES (a cura di G. Garbarini 2018, ListLab, Trento, pp. 52)

This research originally aims at exploring the possibility of interpreting landscape as an ecological infrastructure, an environmental machine able to improve living conditions of inhabitants of the contexts in question (whatever it may be) through a "project" of a landscape smart grid able to organize energy and information flows of communication and culture. The nodes of this network are physical spaces that characterize a new conception of landscape as a network based on the "architecture" of smart grids. These "catalyst"[1] nodes should respond to the needs of a place and its possible changes, applying strategies of visions that show us what we might aspire to.

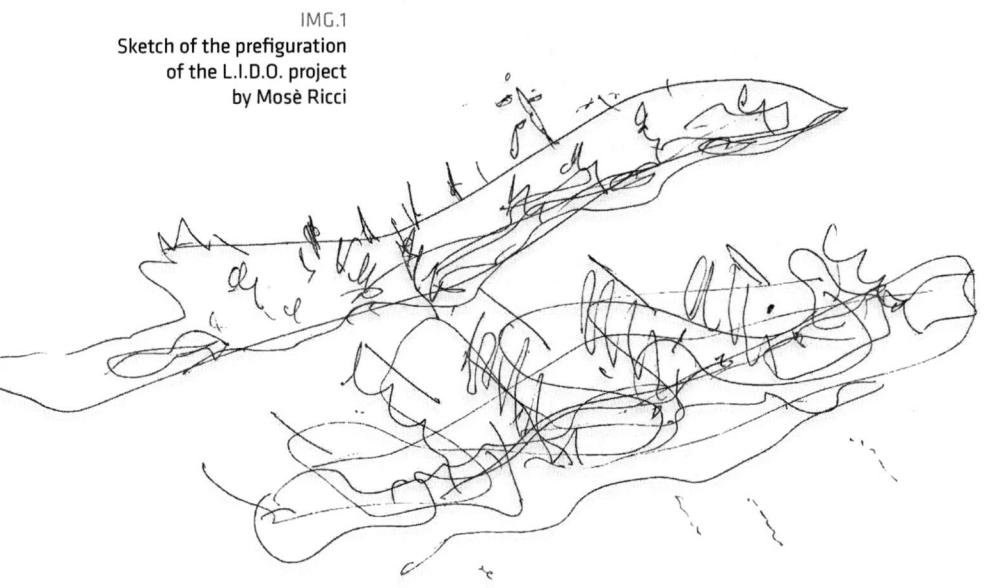

IMG.1
Sketch of the prefiguration of the L.I.D.O. project by Mosè Ricci

FOREWORD

ABSTRACT

Renewable energies play an important role in today's mutable landscapes. The present reseach investigates the on-going sustainable energy transition from a landscape architecture perspective.

A particular attention is given to smart grid systems, which are certainly capable of managing energy flows in a flexible and adaptive manner. The question then arises: **could they also be proactive towards landscapes, as well as flexible and dynamic towards the environment?**

Today, the devices for production and distribution of renewable energy should not only be appropriate to the context but they should adapt to it. Furthermore, these devices cannot only be seen as abstract objects imposed upon a landscape from above but should also encounter local needs.

The present book aims to explore the use of enabling technologies in the landscape context through an inter-scalar approach in a way that makes enabling landscapes pro-actively responding to changes taking place in them.
The main structure and process are based on the architecture of a "micro and macro smart grid", which is generally associated with urban energy grids and districts, but may become a figurative reference for new forms of landscape, such as **"Smart Landscape"**.

The smartness that resides in landscapes, however, cannot only be associated with a technological factor. When we are talking about its (smart) ability, we should think of leveraging various devices and tools and linking them into a broad vision that works by combining points and successful interventions, the way it was already anticipated in the "No-Stop City" of Archizoom. By using smart grid architecture as a network on which to draw a "Smart Landscape" we can exploit its "rigidity" as a potential. It allows us to work no longer with masterplans and plans dictated from above but with a diffused approach (Branzi A., 2006), which in this case works through points and connections triggered by a combination of energy, technology, environmental and social interests.

The concept of network is therefore fundamental: a network could be seen as an image of the contemporary urban condition, where everybody seeks to redefine their perception of cities or urban districts in the age of globalization, networks society, and virtual technologies.

The **output** of the research would be to show how the main strategies of "Smart Landscape" and its development could be applied in different context. The outcomes deriving from the theoretical framework and the identification of reference projects that have already started towards a "smart" context allow us to hypothesize: strategy (Interoperability and Accountability), structure (smart grid), and process; which could be repeatable in other contexts.

The prototype is the island of Venice Lido, to which the concept and structure of the "micro smart grid" would be applied, trying to follow analyses and pilot projects aimed at creating a research project called "L.I.D.O. – Venice: Learning Island Design Opportunities – Venezia. Sustainable scenarios for Venice Lido", already published in 2018 and presented in 2016 at the Venezia Biennial. In this case, the "architectures" of smart grids are able to explore different contextual conditions. As a result, they configure processes / projects that are able to regenerate the existing scenario and project it towards the new ones, thus obtaining new kinds of agricultural landscapes (urban or rural) that have a deep connection with sources of renewable energy.

Smart Landscape is a reflection on development of an urban and landscape design typology linked to the changes brought by the continuous evolution of technologies and the increasingly pressing need for resilience of anthropized contexts, and not only.

Smart Landscape, starting from energy devices for the management and distribution of electricity resources, tends to define a possible vision of landscape, identifying emblematic, successful cases and "smart" projects, which make it evident how the setting up of a unitary process could result in a cellular view of landscape with a reticular function.

INTRODUCTION

Explicit fundamental questions raised by globalization have become a precondition for this research. Today more than ever before, we are speaking about climate changes and their influence on our future often forgetting about the history of our planet and its climate that has always been drastically changing.

For a long time, the climate change subject has been treated in a chaotic way: opposite matters were often mixed, which resulted in creating approximate theories without any valid scientific basis.
This fact is emphasized by the idea that climate changes and temperature swings have always existed and will cyclically appear in the centuries to come. Human actions are affecting their acceleration or accentuation. Therefore, we should try to improve the situation without falling into a disproportionate alarmism.
It is possible to note that various researchers have intended to address the main problem of the twenty-first century, namely that of climate changes, and of how the need to reduce these changes is predominant in our future design challenges.

In the last decade, **Landscape Urbanism** has become the center of debate among practitioners and theorists involved in shaping contemporary city. Advocates of Landscape Urbanism share the idea that traditional dichotomies like that between city and countryside are not able to illustrate contemporary urban realm. Rather, a new urban morphology has evolved, and it calls for new methods and models of approaching cities. Landscape Urbanism suggests re-evaluating landscape in order to develop this approach.
The present research delineates a possibility of drawing inspiration from functions and operational aspects of landscape rather than from its aesthetic qualities. Defining landscape as "landschaft" implies focusing on process, contingency and integration of cultural and natural processes.
While landscape is viewed as a construction and concept open for several interpretations, in the present work preference was given to investigation of the term "Smart Landscape".
The research focuses on a prototype design and intervention model that is configured by adopting the architecture of a

"smart grid" as a design base. This makes it possible to exploit various systematic and technological conceptions in order to define and manage functioning of different landscapes, territorial and urban contexts.

From this emerges an energy-based vision of landscape that could be self-managed in case of deficiencies. This vision could be applied in places where successful projects of intelligent networks or of landscape (urbanism) already exist or are anticipated. Through a strategy (interoperability and accountability), structure (intelligent network) and process we can succeed in delineating within it not only possible visions but also actions, methods and devices that lead to defining an innovative intervention process of managing resources present within landscape or requiring additional interventions.

Similar to city and a territory projects landscape project is in a phase of transition towards energetic, resilient and smart concepts with reference to new research.
The present work is part of contemporary debate on climate change, pollution reduction, "smart technologies", "smart cities", and especially innovative landscape models. It exploits the ability of this debate to "combine" within it a multitude of disciplines from formal to more dogmatic ones, successfully redefining them.
By declining various contents that originate from other disciplines within landscape, the latter acquires its own specificity that grows with time. It changes the application of these disciplines in different contexts, transforms and designs new territories in order to respond to characteristics and necessities of the place in question. It no longer imposes conventional and standardized practices from above.
The main concept developed within this research context is connected to "smartness" consisting of technologies of new and unpublished social and cultural models. Therefore, it seems complicate, at least as regards the objectives of this research, to identify a univocal and somewhat shared definition of "smart" or "smart city", "smart land" and "Smart Landscape".

The **"smart"** approach that should be found in landscapes cannot be linked only to a "pure" technological factor, but rather to the ability to exploit devices and tools by relating them to the context, linking them into a vision that scales landscape, working on specific interventions in response to concrete and imminent needs. This may be done with the help of such an architectural model as energy smart grid, rigid enough in its structural conformation and pre-established rules, which allows a targeted approach due to its characteristics of interoperability and accountability and to a structure consisting of points that are activated as a combination of interests and energy, technological, environmental and social applications.

"Smart Landscape" describes processes of change and transformation of the landscape concept and the way of using a combination of technological innovations and landscape.

The present research is a study of contemporary landscape that addresses problems related to excessive use of natural resources by a rapidly increasing population. The growth processes can hardly be stopped, but they can be addressed and managed by activating processes that redefine sustainability of territories and improve the quality of life of their inhabitants.
Smart Landscape, taking reference and founding its bases both in technological innovation (already present in the contemporary smart city) and in some more advanced conceptions of landscape urbanism, represents an important starting point in a phase of profound changes in the structure of demographic, social, cultural, environmental and economic aspects of landscape. City is seen as an intervention priority within inclusive and sustainable urban growth. Starting with this point, landscape, which is perceived and observed by man, has the ability to function in different contexts in a transdisciplinary way, thus becoming an object of intervention.

It is not the first time in the history of our planet that we are witnessing its population concentrating in urban centers; in the same way, it is not the first time that major climatic changes have put pressure on man and the contexts in which he is settled.

On the other hand, the global nature of these phenomena is unprecedented, the same as the point of "no return", mainly caused by anthropic actions. Thus, various possible interventions should be applied to these actions as they remain the only potentially controllable variable. They can be modified in order to reduce the effects of climate changes on our planet and territories we inhabit by working intelligently and resiliently on the contexts of our interest. In this sense, the strategy of the "European Smart City" model is particularly important, as it is a model of cities that use technologies to promote territorial sustainability. Within this research, we state that such interventions should be focused not on a city or territory but on landscape understood in its broadest meaning.

We have investigated and observed operation and systemization of the smart grid technology and the most avant-garde landscape theories, such as those of Landscape Urbanism. We tried to initiate and encourage integration of environmental, urban and cultural processes. It can be interpreted as an opportunity for creating of new forms of interaction between human activities (their environmental and cultural dimensions) and innovative disciplinary fields, for example, the underlying concept of Smartness. In this logic, technology becomes an instrumental element of Smart Landscape, which adapts to the criteria of sustainability in its various meanings (such as "economy, mobility, environment, people, living, governance"). It derives from a smart city concept, because it is well known that technologies alone are not able to generate wealth, while, if they are part of a strong landscape concept, they can lead to new goals.

By analyzing best practices and taking successful projects as a reference, it was possible to understand in what way to propose technological solutions aimed at promoting sustainable urban development. There are numerous cities (Amsterdam, Stockholm, etc.) that have undertaken this transition process of constructing an intelligent city with the help of digital technologies, but they are mainly metropolis, which, because of their size, can count on availability of resources both at economic and socio-cultural levels. This is why we felt the need to turn our gaze towards landscape and widespread technologies with high potential (like that of smart grids).

Exploring smart grids as an architectural basis for a new spatial planning process through a case study in a deliberately limited and small context (the Island of Venice Lido) has highlighted the fact that design of an intelligent landscape represents a complex operation and must be based on a clear intervention strategy, which can only be addressed with proper planning, focused on the understanding of local specificities, but applied globally.

Identification and description of a new concept of planning and design of urban and territorial contexts, as a result of combining landscape concepts, is performed with the use of a "smart architectural" base composed of interconnected nodes of interest (smart grid). Its main objective is identification of sustainable development processes that propose an innovative prototype. The latter overcomes modernist planning concepts and responds to the real needs in a smart way without unnecessary interventions. This process aims at integrated planning of sustainability of territories and takes into account not only big cities, but also all the minor urban and landscape forms that represent both Italian and European urban structure.

The objective of the research is, therefore, to explore "smart city", "smart land" and Landscape Urbanism strategies in order to rethink the balance between natural and urban environments and technology, aiming at their sustainable development, which intends to pursue the European objectives 2020, 2030 and 2050 to reduce greenhouse gas emissions and to make landscapes resilient to changes.

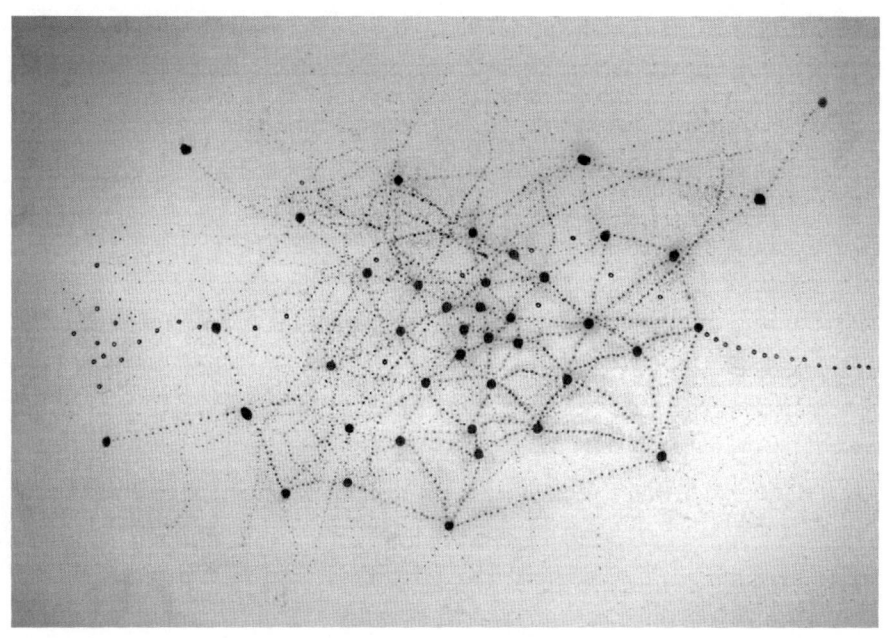

IMG. 2
Emma McNally Imagined Networks

CHAPTER ONE

1

FIVE CONCEPTS OF LANDSCAPE

The list of available literature that addresses the topic of landscape is endless; it allows us to analyze the subject from several points of view. The latter, although in different ways, agree on the fundamental aspect of this "place", which (as mentioned earlier in the quote by M. Jakob) characterizes our era. Landscapes that surround us and "belong" to us cannot be described in a single definition as they are changing in line with the expression and dynamic interaction between natural forces and cultural characteristics of any environment. It is also impossible to circumscribe them in a single disciplinary field, therefore,

> "this historical phase of the research in the field of landscape is characterized by awareness of the limits of reductionism and segregation of knowledge, as well as by diffuse efforts to identify ever-wider approaches to a topic that, day by day, unravels its complex and unmistakable nature through traditional interpretations of ecological or humanistic imprint" (Jakob M., 2009).

Landscape is a result of a consecutive reorganization of the terrain in order to adapt its use and its spatial structure to the changing needs of the society. History has seen many changes in landscape. Sometimes they were devastating, barely leaving any relics, particularly in Europe. Thus, it is normal that today changes are seen as a threat, as a negative evolution because they cause a loss of diversity, consistency and identity. But it is also true that changes can be seen as combined effects of driving forces such as accessibility, urbanization, globalization, culture, society, industrialization, and natural disasters that were remarkably different for each era in the history. They affected and influenced the perception of landscape in different ways. Therefore, by changing the way of using, modeling, studying, protecting and managing landscapes, culture was consequently changed. Diversity and identity of cultural landscapes are therefore central to the discussion, as they show traces of the past of individuals, their roots and identity. Observing the changes allows one to understand the historical evolution of connections between man and nature. This idea is expressed in the European Landscape Convention, which will be used as one of the key items in this chapter.

As shows the evolution of landscape concept in relation to the processes of globalization and industrialization, the latter have created degraded landscapes with an inevitable consequence of loss (for the population inhabiting them) of a territorial reference framework, but have also led to a continuous evolution of this scenario.

Stability, strong attachment to places of origin, and balance between man and territory distinguished archaic societies. However, modernity has introduced de-territorialization, abolition of categories of space and time, and dissolution of distances. The result of this process is homologation and standardization, which also affect landscapes, where the well-established harmony disappears, balance between man and nature fades leaving room only to functional needs. Ultimately, nature and culture are no longer united.

In this connection, it is worth recalling the main conclusions made by E. Turri on the changes and landscape. Turri argues that it could be necessary to declare that landscape is disappearing *"with the process of industrialization and changes related to it, ... as the direct confrontation between man and nature ceased, and a particularly special period came to an end."* We could think that in such a period, the way it happened in the pre-industrial or pre-modern world, man would be able to find the signs of self in the nature. But the signs *"of his way of creating order according to his proper intentions, ... and needs suggested by the culture and society in which he lived and operated"* (Turri E., 2004) might have disappeared from our landscape. Direct connections with nature disappear, leaving space to "manipulated" places: the resulting landscape is altered and no longer recognizable in its original form, although we believe that in this sense it has not disappeared but has evolved.

This altered and "manipulated" landscape does not become describable by a single discipline and definition. But, unlike its primordial meaning, it is almost impossible to study landscape in all its endless facets unless a single choice is made. The theoretical and study choices made inside this research work pursue the goal of creating an evolutionary pathway of the landscape concept in correlation with social, cultural, economic and technological transformations. The time frame of these evolutions is a cultural and social comprehension of landscape. Huge changes that have influenced interpretation and use of landscape are linked to a specific historical period, which becomes a marker of the conceptual change over time, but at the same time, it does not lead to the extinction of the old concept as much as to its evolution.

This paragraph aims to articulate the above-mentioned conceptual evolution, not so much with landscape definitions juxtaposed with adjectives that describe a change or a historical period to which they refer, but with the intention to concretize the evolutionary aspect of landscape that evolvef through such periods as (0.0) Nature, (1.0) Refuge, (2.0) Factory, (3.0) Square, until the present era of Technology / Industry 4.0 and, therefore, Landscape (4.0.) as a Network.

The intent of such parallelism lies in today's historical moment, involving a new technological revolution that is already changing and that will radically modify our way of living, working and relating to each other and to the places we inhabit (Audretsch D.B., 1995).

The first technological revolution as well as the industrial one (1.0) were successful in introducing water and steam as tools of mechanization and production, leading to construction of new urban and landscaping systems. After that, it was electricity (2.0) used to create mass production that introduced numerous changes to the "places", where factories and production chains were located. The third "revolution" (3.0) used electronic and computer technology to automate production, communication and transportation.

And today, where are we? Towards what are we heading? The fourth revolution (4.0), the one in progress, is digital. It started in the middle of the last century and was predominantly characterized by fusion of new technologies, aimed at leveling or cancelling boundaries between different spheres, from physics to digital biology, from projecting disciplines to Landscape Urbanism.

In the nearest future, a foreseeable transformation is going to happen. It will be gradual as well as technological, of a scale and complexity big enough to be considered different from any other experience that has been lived by man before. While not yet knowing exactly when it will happen, one can already clearly state that the response of different scientific communities to this transformation needs to be global and integrated, involving everybody in the world's political system, in the public and private sectors, academic world and civil society. There are three reasons to why today's transformations are not simply an extension of the Third Industrial Revolution, but rather arrival of the Fourth and distinct one, to start with speed, followed by importance and ultimately impact on various systems.

The speed of current innovations has no historical background. When compared to the previous ones, the Fourth Revolution is evolving at an exponential rather than linear pace. In addition, the amplitude and depth of these changes announce transformation of entire production, management, planning and governance systems. Opportunities given to billions of people connected by mobile devices with unprecedented processing power, memory capacity, and access to knowledge are unlimited.

Meanwhile, digital manufacturing technologies are interacting with biological world on a daily basis. Engineers, designers and architects are working with computational design, additive manufacturing, material engineering, and synthetic biology to pave the way for a symbiosis between microorganisms, our bodies, products we consume, and even buildings we live in. Like the revolutions that preceded it, the Fourth Revolution has a potential to increase global income levels and improve quality of life of people around the world. To date, those who have been able to get the most out of this are consumers who can afford access to the digital world.

Technology has made possible the existence of new products and services that increase efficiency and rend our personal life more pleasurable. Ordering a taxi, booking a flight, buying a product, making a payment, listening to music, watching a movie, or playing a game... any of these can now be done remotely. And what will happen in the future?

Will future technological innovation also lead to a miracle on the side of the offer with long-term efficiency gain and productivity?

Probably, the costs of transport and communication will decrease, logistics will become more effective and trade costs will drop. All this will open new markets and stimulate economic growth. So why should this fact not interest the architectural and landscape scientific community? Yes, it should not only interest them but should become central in future projections so that these disciplines could also be able to bring about conceptual and social evolution that will lead to the creation of a clever and avant-garde landscape in the times we are living now, which is Landscape 4.0.

> *"We must see landscape as a delimited territory, defined within a larger reality, that man through his own intervention, triggering a process of choices, transforms into a landscape"*
>
> **Goethe J. W. 1991**

1.1 LANDSCAPE 0.0 NATURE

> "Landscape is a phenomenon, in the Kantian sense of the term, that exists in the moment and in the way we perceive it. If landscape exists only in relation to our experience, then landscape as the thing itself ('dasding an sich') does not exist."
>
> **Chiara Rizzi** - Fourth Landscape

NOTE 8 - IMG. 3
View of "Valdarno"
by Leonardo DaVinci 1473

....

1.

Artification: a kind of realization through art. It takes place in situ, when it is a work of those who intervene directly on the territory and modify it over time following various cultural models, and in visu, when it is a work of painters, writers, photographers, who intervene indirectly on landscape building a model that will influence the collective way of looking at it.

Landscape 0.0 Nature is a mere human interpretation, the starting point in which nature becomes landscape through the eyes of men. The first artificialization of landscape took place with the pictorial representation. Thus, Chiara Rizzi's idea makes us wonder: *if landscape exists only in relation to our perception, how would it be possible to represent and identify it?*

This question becomes central in evaluating the path as well as theoretical and conceptual connections that are created between landscape and the way of representing, understanding and perceiving it. If what Alan Roger describes by the concept of "artialisation"[1] is true, then our perception and therefore understanding of the real landscape would be guided or conditioned by an imaginative, unrealistic landscape, or at least the one reinterpreted by its proponent. The natural place of an individual's habitat is perceived in an aesthetic way only through landscape, which can be changeable, and which allows us to determine its nature through art. Landscape, once again quoting Alan Roger, would be a cultural invention, which in no

1. FIVE CONCEPTS OF LANDSCAPE

way could be reduced to its physical dimension, but in order to become what it is in the life and perception of man, it always needs a metamorphosis, which appears essentially in the reality of art (Roger A., 1997). Therefore, there is a place that represents the zero level from which landscape is modeled, and this modeling takes place through a 'metaphysical' process, which does not belong to nature.

Landscape, unlike a place, is not composed of objects but it is only a way of seeing and representing these objects in the world (Farinelli F., 2003). In geography, this, thanks to Alexander von Humboldt[2], is outlined with three stages of knowledge: the first one concerns ideas, which arise in the soul of man as an original manifestation, a primordial feeling that is manifested as an answer to the grandeur and beauty of nature; the second one is born from examining nature, disintegrating its sentimental totality and translating it into scientific terms followed by an emotionless dissection of its components; the third one being the closing of a conceptual circle is the return to totality that rebuilds an overall picture and refers to being one with an interdependence of the elements previously decomposed. Theoretical transition from pictorial to scientifically relevant landscape takes place thanks o generalization, which is proposed by art.

If, as a result of admiring a picturesque view, we can formulate scientific views on landscape, then they should represent a clear view of the starting point they deal with. Taking into account that landscape would not exist if not interpreted by a human being, we can say that the concept of landscape started to appear between the 13th and 14th centuries (at least from the pictorial point of view) in the major schools of Northern Europe including the 16th century Flemish School, the 17th Dutch, the 18th – 19th English, and the 19th French.

However, if one wants to go back to the oldest (before Lorenzetti[3]) insight and description of landscape in its pictorial context, one should turn to the field of literature. That is where we find Francesco Petrarca[4], who, in 1336, during his journey to the summit of Ventoux in the South Western Alps, tells his correspondent about what he sees from the summit of the promontory, with a new perspective of what will then be considered as 'landscape' looking away from the limits introduced by religion and the banal views that people of the internal parts of the country used to have (Küster H., 2010).

2.

Alexander von Humboldt – naturalist, explorer and botanist. In the first half of the last century, he succeeded in convincing the European and American bourgeoisie to study the sciences of nature. Thanks to him, in fact, the concept of landscape changed drastically from aesthetic to scientific, switching from pictorial and poetic knowledge to the geognostic description of the world. It got loaded with a completely new and literally revolutionary meaning from the point of view of the history of knowledge.

3.

Title of the work: Effetti del buon governo in città e in campagna. Author: Ambrogio Lorenzetti. Dating: 1338-1339. Location: Siena, Palazzo Pubblico, Sala della Pace.

4.

Francesco Petrarch, "Ascent of Mont Ventoux" (Familiares, IV, 1). The letter, addressed to the Augustinian friar Dionysius of Borgo S. Sepulco, the spiritual adviser of the author, tells the story of a trip taken by Francesco and his brother Gherardo to Mont Ventoux.

We assume that independently of the form of landscape to be read or interpreted it always remains what Plato's man[5] perceived in his environment and with what he related. In any way it is meant to be interpreted or represented, it always consists of a connection between elements of nature and culture enclosed in a single compressive vision of a place.

To support this statement, one can only begin with nature, which makes part of landscape and thus becomes a fundamental passage that has led nature itself to be recognized and counted within the vast universe of landscape.

To do this, one must not forget that the perception of landscape in the past was not as immediate as it is today. Man was born and evolved within nature and was part of it, nature was his home, his place of "work", this way man himself was nature.

In the Middle Ages, nature began to be perceived as a place of loss, dominated and controlled by the devil in person; it was feared by man, above all if seen through a religious filter that considered it to be a source of distraction both on the path of approaching the divine and in the construction of a dichotomous binomial where cities (home of civilization and man) were opposed to nature, the center of chaos.

In the 14th century, something begins to change and nature starts to be seen from a new perspective: it is enough to think of what was said by Petrarch in front of the Lorenzetti fresco in the Palazzo Pubblico of Siena, where for the first time in the pictorial history he portrayed the end of the division between what was created by man (city) and what surrounded it ("wild nature"). There is a perception of world dissolution in a "liquid" society where nature itself has also returned to the center of the discourse.

From the urban point of view, a perception of man-made nature is present in the famous Lorenzetti fresco for the first time in Western post-antique art. It represents a clear separation between city and countryside but at the same time symbiosis between the two, symbolized by exchange at their borders. The predominant aesthetic aspect is represented by the control that city exerts on countryside without even glimpsing a look at nature as such (absence of a horizon) but seeing it only as a political or agrarian territory. Subsequently, numerous variants of this connection were born[6]: already in the 15th century, Leonardo da Vinci[7] spoke of "towns" with reference to visual

---

5.

Man, in his original state, inserted in nature, without any specialization, creates the first place to live in: a landscape of the Great Mother, where he begins his activity of Demiurge, as Plato said, of the creator of the world, who is present in the genesis and the Old Testament.

perception and consequent representation of natural and anthropic environments: "The Aspects of Cities", "The Way of Forming Cities", "The Cities made in the figuration of Verno", "Retreat Sites and Cities" (Buccaro A., 2011). There are several ways to point out the techniques of new landscape paintings by those who inaugurated them in 1473 with the famous view of "Valdarno" towards the bog of Fucecchio commemorated from Monte Albano[8], which is kept in the Uffizi: in his paintings, the Tuscan genius exalted some treats linked to the vision of the territory by filtering natural factors such as vegetation, light, atmosphere, distance, but also signs of presence of man and his various creations. But even if these early studies were not considered landscape (or probably they were, but only later), we can notice that from that moment on the logic of the artistic representation changed radically.

From there on started a proliferation of figures that contributed to giving landscape a new meaning and freeing it from any religious reference. The new era was launched by Ambrogio Lorenzetti's fresco "Effect of Good Government on City and Country" or by the famous opera "City by the sea"[9]. Starting from the works of Lorenzetti, Pisanello, and Leonardo da Vinci, the artistic interest in Europe was directed to the representation of nature (though later it was for a long time used only as a reference to adjust the decoration). While in Northern Europe we could observe a union of representation of soul, plants and cities that lead to the discovery of landscape.

Of great importance are the works of artists such as Claude Lorrain[10], where the figures were not a guide but a way of bringing man closer to the nature, which was tamed and no longer caused any fear. Caspar David Friedrich[11], on the other hand, reaches a further level in representing nature, the landscaping, in his works using a symbolic religious language with the presence of a figure always turned with his back to the viewer at a distant point of the canvas, so far away that he is immediately related to nature. Remarkable is the invention of the "window", or that of the view that opens to the extreme through a window and is represented in the picture, where it insulates and encloses the countryside transforming it into landscape.

Moving on to the representation of the sublime, in which a "scary" spectacle is transposed on canvas, this spectacle however intended to provoke pleasure in the sublime enjoyment

9.

IMG. 4
"City by the sea", around 1340. Tempera on board, 23x32 cm. Siena. By Jean Starobinski

10.

IMG. 5
Imaginary view of Tívoli, 1642. By Claude Lorrain

11.

IMG. 6
"The wanderer above the Sea of Fog", 1818. By Caspar David Friedrich.

of fear. This way, beautiful brings pleasure and sublime brings enjoyment, thus allowing us to get strength in nature, which is always transformed into landscape through its representation. Thus, there is a prospective separation from the Middle Ages; nature is desecrated and simple countryside becomes landscape (pays-sage or wise country) through art, insertion and creation of certain elements in it, uniting the theory that nature is indeterminate and never receives its determination if not in art. It is interesting to note that the word "landscape" in different European languages means both the representation and the object itself. The French word "paysage", used for the first time in the 15th century, is a neologism; it encloses the noun "pays" (country) and the suffix "-age" (together, totality, globality). "Paysage" as well as the Italian "paesaggio" are new expressions, created to describe an original reality, while in English ("landscape"), German and Dutch, even if they subsequently assume an aesthetic value, they will continue to describe a territory. The French as well as Italian terms refer to a pictorial representation of nature rather than to a represented object. They enrich the (new) meaning of the word "country" with the help of a suffix, which in fact expands correlation between object and subject, referring to a view and only later to a representation of the object through the experience we have of it. At the same time, the proliferation of images of "landscape" is not considered to be an absolute good, as M. Jakob claims in his book "The iconography of landscape"[12], where he argues that "landscape resembles a sparkling text on a screen, where its meaning can be created, amplified, modified, developed and finally canceled by simply pressing a button". Precisely through this thought and with a series of paradoxes, M. Jakob shows the difficulty of elaborating a unitary theory around the topic of landscape. Representation is one of these paradoxes, expressed in how we can fix a phenomenon that by its nature has a floating and open identity.

Finally, M. Jakob identifies and establishes a formula consisting of three essential preconditions:

$$P = S + N$$

That is, landscape requires the presence of a subject, nature, and connection between the two elements to become definable (Jakob M., 2009). This formula represents a definite point within the debate on landscape and the result of this debate, which has been going on for about two centuries.

12.
IMG. 7
Book : "The iconography of landscape"

1.2 LANDSCAPE 1.0 REFUGE

"Utopia (from ancient Greek): ou-topos (non-place) and eu-topos (happy place). In the long rational construction of an imaginary society it opens a new vision of a better world; The Utopian Island is therefore literally a 'non-existent happy place'".

Emanuele Sommariva - *Cre(eat)ing City*

An image acts as a powerful amplifier that allows one to go beyond the primary or immediate effects of perception, giving space to almost unlimited visual associations. The 17th century became a century of "Naturphilosophie"[13] (Zavatta B., 2005), the philosophy of man's glance on nature.

The following century, on the other hand, was characterized by man's transformation of nature: it was first marked by large-scale changes in European territory and, later, in various regions of the world. This transformations indirectly led to a change in perception of nature, maintaining natural characteristics expressed previously but evolving in contexts in which it was necessary.

It is in that moment that towns modified by man in a pre-industrial way replaced nature, but since nature itself was a part of human soul, it was maintained within towns, preserved and reproduced as a safe place, i.e. landscape (1.0) as a refuge where to escape to from what at the same time was destroying and creating it, i.e. human evolution. In the course of history, the wounds that industries inflicted on nature led

IMG. 8
Aerial View of central Park. Image Courtesy: Will Codwell

[13.]
Naturphilosophie is a philosophical theory of nature very close to the thought of German idealism and Romanticism. Going beyond the Kantian definition, in Germany Naturphilosophie becomes a movement of thought, which attempts to interpret nature in a parascientific way, denying the assumptions of traditional science,

to a radical dualism characterized on the one hand by a boom of the "old" landscape, preserved and reproduced in the form of gardens[14], and on the other, by creation of "non-places"[15] (Augé M., 2009), residues of mechanical and technologic civilization.

Creation of the concept of landscape as an independent subject born during the Renaissance consolidated the impossibility of representing "culture" and "nature", which led us to representing landscape understood as a "natural scene, mediated by culture" (Mitchell W. J. T. 2005)[16], which subsequently became a response to the First Industrial Revolution (1.0).

This evolution triggered remarkable transformation in anthropic territories, changing them into unpleasant and polluted places unable to aspire to an acceptable standard of quality of life of its inhabitants. The latter turned their gaze towards untouched nature as a refuge. At the same time it was considered too wild, which forced architects and urbanists to propose alternative settlement models. This change of perspective stemmed from the collapse of the landscape idea, which had united man with nature, but at the same time was disrupted by human actions, which led to profound changes in natural scenarios.

Man, by his constitution, needs bounds. He acts in the world by ordering it, constituting a "nomos" that, in turn, determines specific reference horizons within which he finds himself. Nature made significant becomes culture and subsequently becomes confined and enclosed in places of beauty suitable for reconnecting man with itself. But if it is through art that nature acquires beauty and this way (in the era of hyper-industrialization) is no longer practicable being a wild, contaminated, defaced site, it can become a place where nature and art coexist, where the mythical universal project of liberated nature and the realm of beauty become reality.

This can only happen in gardens, which are products of nostalgia and mirror image in comprehension of the beautiful, where intimacy, fantasy, utopia, escape into idyll, and disdain of reality are projected. For instance, in the sumptuous Renaissance gardens, iconographic programs were created. They combined an Arcadian ideal with the mystery of nature's mythology, and the figures that appeared in caves are derived from the ancient tradition.

---

especially Newtonian one, that is: mechanichal approach and materialism. Even if it was never accepted by scientists, Naturphilosophie certainly had the merit of generating great interest in new sciences of the period (electrology, magnetism, biology) among philosophers, writers, poets, and scholars of various disciplines.

14.
Intended as a closed space within the urban context but separated from the latter, seen as a wild, but at the same time tamed space, which represents the concentration and expansion of what lies outside of the cities.

15.
"non-places" are all the structures needed to accelerate circulation of people and goods, means of transport, large shopping malls, factories, shopping centers ... all the spaces, where

The Renaissance aspired to restoration of the pre-medieval world and, through arts, created a universal momentum for the future, to which (compared to antiquity) a much more important role was assigned. In this age man was able to dominate nature, and this was the result of rational thought extended to political and social relations with trust in progress, which was destined to become the driving force of modern times.

With the Platonic Academy[17] formulated a theory that took a decisive step in the evolution of garden art, so that it became more closely related to knowledge and culture than to the direct pleasure given by nature itself, and thus turned into artwork. Evolution of the gardens of Tuscan villas in the mid-15th century was presented by the pictorial cycle of Giusto Utens, a painter of Flemish origins, who documented the entire patrimony of the Medici villas[18] (now preserved in the Historical Museum of Florence). The images of the gardens of the fourteen villas, painted around 1598 in a bird's eye view, provide us with a good deal of information. In the Renaissance context, they do not have a definite shape of a garden, as rural buildings are surrounded by grids and numerous orchards, olive groves, fields and meadows; these images still reflect the economic aspect of a villa, while the decorative aspect gains some importance only in the immediate vicinity to the house, for example, in the form of "secret gardens" enclosed within walls. A secret garden, derived from medieval "hortus conclusus", was a closed space filled with flowers and aromatic herbs like marjoram and basil, grown in terracotta pots and arranged around the lawn, usually adorned with cleverly cut pottery and boxwood figures. The avenues had to be bordered with symmetrically arranged flowery trees, such as pomegranates and cherry trees, surrounded by roses, bushes and aromatic evergreens forming geometric figures. Pergolas of vines supported by marble columns were to alternate with labyrinths and bushes of juniper, myrtle, oak and cypresses covered with ivy. There were fountains surrounded by flower vases and amphorae with flowers. "Labyrinth" is a prehistoric word of uncertain etymology used by classical authors to name complex architectural works, and especially the Cretan palace of Minos, designed by Daedalus. In the end, there remains a concept,

17.
In 1459, the "New Platonic Academy" was founded by the will of Cosimo dei Medici. It was also called the "Neoplatonic Academy". The cultural context in which the Academy was operating was then strongly marked by Platonism, reborn in Italy towards the end of the 15th century, during humanism. Scholars and artists frequented the place assiduously: Giovanni Pico of the Counts of Mirandola

18.
IMM. 9-10-11
The Villa di Castello in a lunette by Giusto Utens.

La villa nel 1599, Lunetta di Giusto Utens.

Palazzo Pitti, lunetta di Giusto Utens del 1599

perhaps not original, of complexity, tangled with structures and passages, which also designates a decorative motif (solar symbol) similar to the meander, but radial instead of longitudinal. When accessing a labyrinth, the median axis is already out of sight and in this way the upward orientation is prevented; the sensation is that of the body left alone, where the inevitable happens, and one is lost in the absence of an implicit way out of over evaluation of self and its inadequacy. Man does not rush there because of a strange will, but for his own action and decision-making. The rigorous geometry refers to the fact that only reasoning can indicate the way out.

Garden thus became a tool of control and a protected place, where it was possible to recreate natural, idyllic landscape, in which man of the first industrial revolution was seeking to rediscover the duality with the nature that seemed to have been lost in that century.

As briefly mentioned above, to get that far an individual who resides in a landscape is never satisfied, he must and will be guided by architects and urbanists during his rediscovery of nature, and the latter will interpret this primordial need of returning to nature and therefore finding a refuge from industries in the utopia of a "garden city"[19].

One of the greatest exponents of this utopian movement considered in this chapter is Ebenezer Howard[20]. With his works, the urban model begins to be conceived not as a project but as a set of principles, norms and procedures; it precedes and facilitates social reforms. The author's proposal (1898) set the goal of decongesting a historic city, while planning and managing expansion through decentralization of the population into a newly-formed city called "garden city". Starting from the finding that both city and countryside have positive and negative aspects for the existence of man, Howard in a third entity – the city-countryside – locates the ideal solution that encompasses benefits of both urban and wild life while allowing the elimination of characteristics unfavorable for human existence. In the Howard's spatial plan, a system of satellite towns immersed in the green, sufficiently spaced to avoid welding, is located in the heart of the central city. The "garden cities" are thought to be self-sufficient, with 32,000 inhabitants. They have radiocentric structure and

---

and Concordia, Angelo Poliziano, Sandro Botticelli, Cristoforo Landino, Leon Battista Alberti, and this is anticipated in their works.

19.

Garden city movement was born at the end of the 19th century with the aim of decongsting big cities through decentralization of their population by moving it to satellite cities immersed in the green.

20.

In his book "Garden Cities of Tomorrow" (1898-1902), Howard sets up an adaptation model adjustable to various territorial contexts, and is therefore taken into account in the discussion on the topic of garden cities.

---

1. FIVE CONCEPTS OF LANDSCAPE

are connected to each other by a road system, a network of channels and a rail[21].

In the middle of the city, there is a garden. The heart of this utopia is based on a feeling of belonging to nature, which is an inseparable part of man (without which, as we know from a book by Fabietti U. et al., 2000, there is no landscape). Man in this context is in a search of connection with landscape. The garden is an image of a utopian desire (Fabietti U. et al., 2000). The biblical story of paradise symbolically describes an original and natural life, and promises, for the end of time, a realm of perfect harmony. The Garden of Eden becomes over time a model, a prime form of large gardens throughout the centuries, because it exists as a myth mediated by literature, through the dream of humanity in a life characterized by perennial happiness.

Through art, nature acquires beauty, and it is no longer impractical and dangerous as a wild place, but it becomes a space where nature and art are united, where the mythical universal project of a liberated nature and the realm of beauty becomes reality. Gardens are a product of nostalgia and mirror image in the dimension of the beautiful, where intimacy and fantasy, utopia, escape into idyll, and disdain of reality are projected. Thus, Landscape (1.0) is born, where, as a result of creation of an ornamental garden, an individual finds a place to rediscover the pleasures of the senses, contemplation, rest from work. Within the unstoppable evolution of industrialized and avant-garde man, it is the evolutionary passage that follows the primordial conception of landscape as nature, consequently leading to human evolution.

21.

IMM. 12-13
Garden Cities by
Ebenezer Howard.

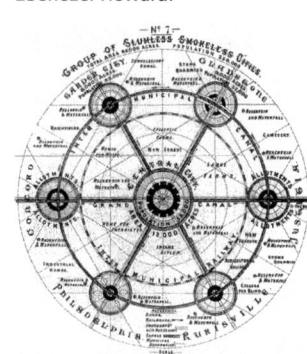

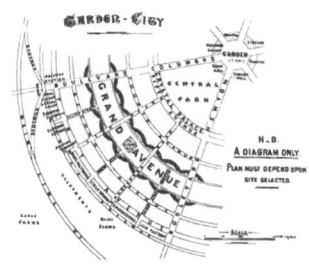

1.3 LANDSCAPE 2.0 FACTORY

IMM. 14
Oldham from Glodwick by James Howe Carse (1831).

"Productive surface is soil that has the ability to produce something. In other words, it has intangible and positive by-products: energy and biotic or abiotic components. Productive surface depends on a profound understanding of the context, climate and natural processes. It can operate on the architectural, regional or intermediate scales, due to its logic of network and scalability. (...) A productive area recognizes and capitalizes its innate potential for seasonal or cyclical performance. It is dynamic and responsive, but also usable and tangible. (...) How could production surfaces generate new economies, programs, typologies and public areas?"

M. White, M. Przybylski - *On Farming: Bracket*

22.
The artistic current of Romanticism of 17th-18th centuries is an expression of acceptance and exaltation of the typical features of human consciousness and behavior such as melancholy, irrationality, doubt, despair, dissatisfaction with the repetition of 'normal' life, desire to be absorbed by the forces of nature until the annulment of self.

After having explored pictorial and etymological meanings of the term "landscape", this research makes an effort to understand to which point this discipline was linked to aesthetic experiences, to its nature and its processes in the past (this idea is related to the romantic artistic flow[22] of, just to name a few, Baudelaire, Friedrich and Goethe). These artists circumscribed the value of landscape to its mere contemplation and

– even more – to projection of it to be admired... or to reproduction of it in order to create a refuge from modernity. These views on landscape are undoubtedly reductive as soon as they do not allow us to grasp its true essence and abundance. Within this abundance, we find some anthropic aspects that influence the evolution of landscape itself. Indeed, the advent of the second industrial revolution (2.0) led man to work on landscape, not only constructing factories for massive production, but turning landscape itself into a factory intended to satisfy his needs.

We can generally state that most "spectators of landscape" agree that the snowy summits of the Dolomites are a pleasure for the eye, that contemplating the sunset of the Venice lagoon arouses a set of unique emotions in a human soul, and that the landscape of the Ligurian coast between Monterosso and Tellaro[23] offers breathtaking natural sights. These three panoramas, albeit marked by the presence of "wild" nature, are the result of man-made alterations of a land due to economic, productive, cultural and social needs. They are rich in history of a group of people, a community that shares interests, language, customs and traditions.

Territorial changes due to economic reasons are the ones that most contributed to the transformation of landscape in time. We must not forget that the above described landscapes are first of all productive surfaces. It is enough to think of the lagoon fishing, the terraces of the Ligurian coast or the importance of vineyards and apple fields in Trentino-Alto Adige, which already existed in the past. Economic reasons have always been the ones that were pushing the inhabitants to modify their land through agriculture, timber harvesting, construction, energy exploitation, etc. ... man considers the territory as a surface capable of generating certain profits or as one of the actors of production. It is properly defined by Mason White in the quotation at the beginning of this paragraph, which ends with a question to reflect on. White closes the article (White M., Przybylski M., 2010) with a question: *"(...) How could production surfaces generate new economies, programs, typologies and public areas?"*

The territory is resubmitted as an economic asset to be "used" and not just as a diary of signs of man's presence in nature. Exploitation of the resources should not be under-

23.
The province of La Spezia, Italy, includes the Gulf of La Spezia and the Cinque terre (Unesco World Heritage).

stood according to the mechanism of how much the land produces, but rather from the point of view of transformation of the goods. Landscape thus is no longer contemplated as a treasure chest of natural beauties or traditions, but within this vision it is also an active subject of life of its inhabitants and economy of the territory. However, it is to be understood that a productive landscape cannot be adapted to any type of economy. It becomes necessary to identify sustainable activities both from natural and economic points of view, such as those of the agricultural sector and services, which, thanks to the use of innovative technologies and their ability to exploit natural resources, can function better without altering the balance of an ecosystem.

Like castles or monasteries, which were foundation centers of a multitude of villages and towns in Middle Ages, the same way factories, starting from the late modern period, were configured as a new type of territorial emergency able of triggering mechanisms of progressive aggregation of settlements aimed at creating true workers villages. The origin of this type of settlement is a gradual transition to the centered manufacturing and the consequent functional integration of workplaces with the help of infrastructures and equipment essential for the community life of workers.

It can be stated that over the years landscape shifted from its more classic and static meaning of "heritage" to a new one that identifies it as a productive surface capable of enhancing and exploiting natural resources inherent in a given context. During the Modernist period, a deep and silent transition of landscapes took place, transforming them into production landscapes that, in relation to the proven needs or their location, exploited natural and spatial contexts to become productive surfaces. Literally, according to M. White, a productive area corresponded and referred to the capacity of a designed and anthropogenic surface to generate functional components, such as agriculture, renewable energy systems, water collection systems, etc. (White M., 2015). In fact, a productive surface can depend on the in-depth understanding of the context, climate and natural processes, and it can work both on a regional scale and at a structural level due to its system and logics of a network.

If in the previous paragraph the utopia of a "garden city" guided us to the interpretation of landscape as a refuge from industrialization, in this paragraph Tony Garnier's Industrial City will lead us to explore the transformation of the concept of productive surface/landscape towards a more, in our opinion, comprehensive landscape concept (2.0) as factory. Howard's theory and design of a "garden city", structured as an ideal alternative to the landscape morphology that was to be developed at the end of the 19th century, could be of great help to us. Garnier develops his Utopia of the Industrial City by extending the desire for utopia, and therefore finding a "happy place" for modern-day citizens who are progressively moving towards a vanguard territory, and thanks to industries are growing up together with technological innovations. The author will study and elaborate a careful proposal based on a contemporary ambition to incorporate productive zones into residential ones. Unlike the "garden city" designed by Howard, Garnier's vision anticipates prospects of growth and evolution through a linear expansion plan. Conceived and strategically programmed to absorb industry's productive capabilities, such as energy, food and strong economy, it automatically transforms city and surrounding landscape into a production site, therefore into a Factory.

Connotation of this kind of landscape coincides (not accidentally) with introduction of electricity, during the so-called second industrial revolution (2.0) and hence with evolution of the world's manufacturing industry, which starts using electrical energy to create mass production. Electricity brought various changes to "places" where factories, production chains, and cities were extending remarkably. This way starts a progressive transformation of landscape, within which these settlements were located, and landscape itself becomes a source of production of goods suitable for industrial expansion and progress, where the industries are located and the land is cultivated to produce food, coal is extracted to provide electricity, transport systems are set up to facilitate communications etc.

Landscape as Factory is also a "welcoming context" (Palace, 2002), a scenario of representation of an intertwined connection between point, linear, and area elements, synthetic devices, a range of specific forms of production, usage, and connection systems.

The conception of a Landscape (2.0) Factory, its identification and spatial location are accomplished on different levels that involve various disciplinary sectors: urban, environmental, socio-economic and legislative. This landscape is created by intertwining elements of various nature, which are put into a system and transformed into multifunctional agricultural spaces. Therefore, landscape resembles a factory, as it produces various goods and values, such as agriculture, zootechnics, fish farming, timber production (where water is used as a means of locomotion of machinery and as a source of electricity production), and, finally, energy production systems (wind and sun power, hydropower, Earth's internal heat).

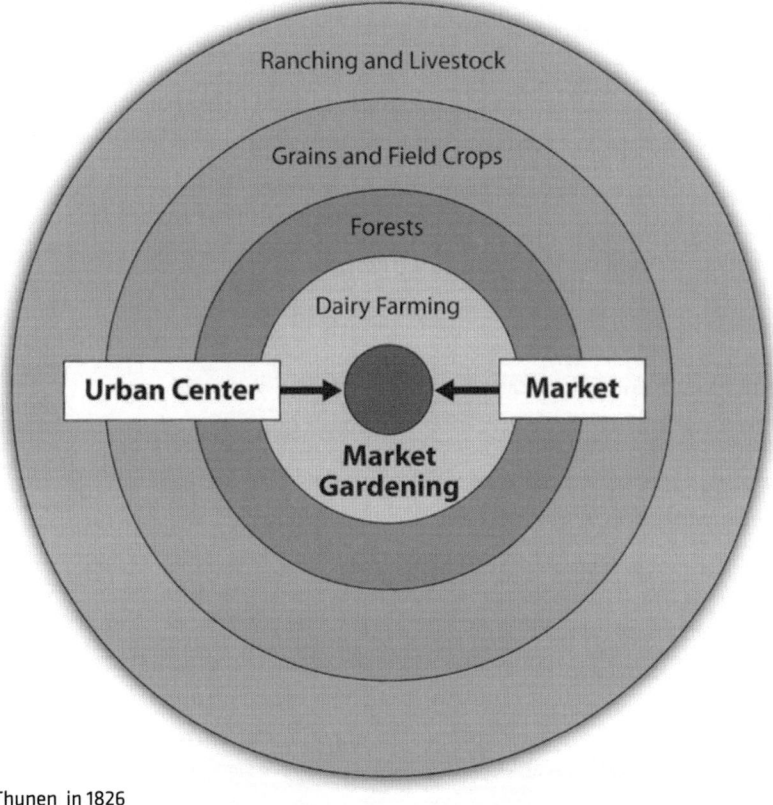

IMG.15
J.H. Von Thunen in 1826

1.4 LANDSCAPE 3.0 SQUARE

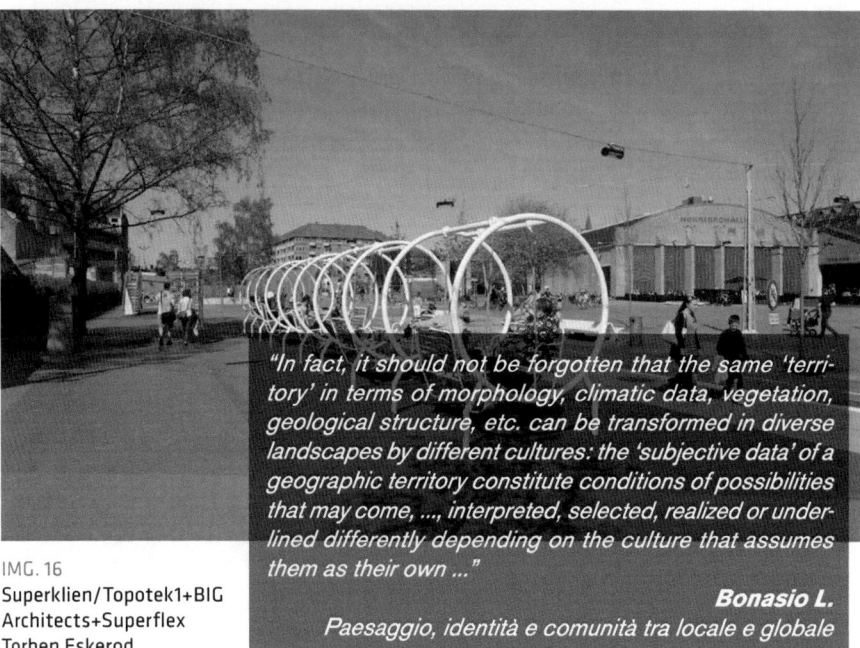

"In fact, it should not be forgotten that the same 'territory' in terms of morphology, climatic data, vegetation, geological structure, etc. can be transformed in diverse landscapes by different cultures: the 'subjective data' of a geographic territory constitute conditions of possibilities that may come, ..., interpreted, selected, realized or underlined differently depending on the culture that assumes them as their own ..."

Bonasio L.
Paesaggio, identità e comunità tra locale e globale

IMG. 16
Superklien/Topotek1+BIG
Architects+Superflex
Torben Eskerod

After a period strictly linked to the needs of industrial and economic growth of civilizations, landscape was rediscovered through the concept of landscape 3.0 (landscape as square) intended as a fundamental place for meeting and exchange, where culture and history, symbols and traditions intertwine. Landscape as square indicates the importance of both these contexts as "vital centers of a city and community, sort of stages for identity and sense of belonging to a society that allows daily manifestations of community"[24].

L. Bonasio, professor of geophilosophy[25] of the University of Pavia, explains that already Descartes[26] stated that it was necessary to model nature in order to make it more accessible for humans; therefore, it was necessary to create landscaping facilities that could guarantee an individual an easier access

24.
Guarente Sergio, Il significato politico filosofico della piazza nella storia d'Italia, http://bilinguescutari.altervista.org

25.
Theoretical guidelines and studies intended as transdisciplinary knowledge involved in collecting and comparing perspectives of different matrixes deriving from geography, philosophy, aesthetics, anthropology and architecture. At the center of interest is the topic of plurality of places on the Earth compared to the increasing similarity of techniques in a globalized world. Treccani – 21st Century Script (2012).

26.
See Bonasio, L. Schmidt M. di Friedberg, 1999, L'anima del paesaggio tra estetica e geografia, Mimesis, Milano.

to "places" (thus creating large plains, straight avenues and eliminating differences). In doing so, life of men was facilitated, but became inexorably destined to lose some of the peculiarities of the contexts where such facilitations were applied. Therefore it is useful to ask ourselves: how can one act to modify the planet for humans without depriving it of its original values?

To do this one has to talk about **LANDSCAPE**.

First of all, one has to consider a person who is a stranger to landscape, an observer of it, in whom the place where he or she is located causes emotions that differ depending on the context. Therefore, in proportion to the emotions that can be experienced, there are different landscapes, each with its uniqueness derived from various factors, some more visible than others. It is still to be remembered that all the diverse landscapes with all their contrasts and differences make part of one and the same planet. As a result, one should never think of a landscape as if it were enclosed in a frame, but should observe it in more general and broad terms. As mentioned above, one must look at landscapes as "vital center ... the stage for identity and sense of belonging to a society".

Therefore, it becomes evident that not all places possess the same aesthetic qualities, but at least in general they are able to express different cultural (mostly local) identities, which are worth being transmitted. It should not be forgotten that one and the same territory can be conceived and modernized into different landscapes by different cultures. Landscapes always being a cultural type of construction should not be confused with or reduced to the mere name of environment, which represents only natural and ecological conditions. Landscape in its total understanding is a container of several symbolic consistencies, of identity and aesthetics, which represent communities that lived and inhabited it, becoming a true identifying place recognizable by its physiognomy and sense of belonging. This extension of the concept of landscape to all places, which surpasses the opinion-based and visibilistic notion that represented a strong aesthetic reduction, opens to a concept of landscape as a place and insuppressible expression of cultural identity. It is clearly expressed in the European Landscape Convention (2000) of the European Council, where the difference between various categories of landscapes (exceptional, degraded, common) entails disconnection of the very concept of landscape from a purely constituent principle, normally adopted for protection of exceptional landscapes, but rarely applicable to others, to projects of improvement or management of different places, including common or productive ones.

There is therefore an explicit and strong appeal to Article 5 of the General Measures of the European Landscape Convention, which recognizes landscape as a cultural identity[27].

This important rethinking of landscape idea allows one to escape from the need of accepting polar alternatives, which would be freezing and museing of a territory on the one hand, and freedom of manipulating with it on the other. This way, it becomes possible to finally recognize unity of individual landscapes and not their logical decomposition. Although using this approach one should never lose vision of the fact that the sense of a place or landscape physiognomy requires unified (but not unpractical and unrealistic) visions and management in order to be maintained.

The cornerstone of this approach lies in assuming a cultural character of landscape and historical, almost indissoluble intertwining of environment and man, in exploiting the issues of a territory that is suspended between city and countryside, and in dealing with them in an extremely specific manner. The latter lies on the border between architecture and landscape and becomes a proof of the need to introduce continuous and indispensable inter-disciplinarity between the fields of action. The effectiveness of landscape disciplines in addressing social issues of conviviality and urbanism is perhaps connected to their ability to work with temporary dynamics of nature, to cross all stages of projecting and to know how to connect a territory with the physical construction of urban space (Palazzo E., 2010).

The use of the concept of landscape is not univocal because it is subordinate to the way in which landscape is lived or observed over time. It does not represent only a physical-environmental or political-social unity, but it has a living nature, "in which civilization reflects and recognizes itself, identifying itself in its forms" (Assunto R., 1973). It is a hybrid system that

---

27.
"Each Party undertakes: • to recognize landscapes in law as an essential component of people's surroundings, an expression of the diversity of their shared cultural and natural heritage, and a foundation of their identity; • to establish and implement landscape policies aimed at landscape protection, management and planning through the adoption of the specific measures set out in Article 6; • to establish procedures for the participation of the general public, local and regional authorities, and other parties with an interest in the definition and implementation of the landscape policies mentioned in paragraph b above; • to integrate landscape into its regional and town planning policies and in its cultural, environmental, agricultural, social and economic policies, as well as in any other policies with possible direct or indirect impact on landscape." (Council of Europe, 2000)

IMG. 17
Moving Forest
by NL Architects

....
28.

Territory, while the latter is home to natural and anthropic dynamics, landscape remains a perception of such dynamics by the population that inhabits it, and its existence is not possible if not through the gaze of this community. For this reason, for the ELC, there is a close link between landscape and subjects (individual and collective) that relate to it, finding in it a background to their own existence.

swings from the highest forms of urbanization or anthropization to the virgin and unpolluted spaces of nature. Connection between city and landscape is a topic that is central to urban planning in the evolution of "green" as urban material and of progressive conceptualization of the landscape idea as a 'square' (a meeting place and a commonplace for individuals). And this is particularly evident in the ELC, which defines it as "a certain part of territory, as perceived by inhabitants, whose nature derives from the action of natural and / or human factors and their interrelations." (ELC, 2000 art.1 (a))

The European Landscape Convention issued in 2000 in Florence by member states of the Council of Europe was officially adapted in Italy in 2006. It was drafted at the time of the Third Industrial Revolution (3.0), when internet and new communication technologies revolutionized the social structure and consequently the use of space around us, giving rise to a "widespread city". This structural transformation of urban space and of most of the 21st century cities has led to production of hybrid territories where buildings, open spaces, infrastructures, and natural elements alternate in a porous tissue, where the boundaries between city and countryside do not exist anymore.

In the historical context, in which it was "born", the ELC marks a turning point in the landscape matter. In fact, this document makes it evident from the very definition of the concept, that it *"does not consider landscape as a purely objective phenomenon (such as territory or environment), or purely subjective one (the notorious landscape as a state of mind) but rather constituting the interaction between the two sides"* (D'Angelo, 2009, p. 35). As anticipated by D'Angelo's words, the accent put by the Convention on the perceptual dimension inherent in the concept of landscape has as the first consequence the fact that it made clear the idea that landscape differs from other "geographical objects", territory in the first place[28].

But this link between population and landscape is not limited to what is stated above: landscape is not only an object of man's perception and a background for his actions, but it is a living reality that is constantly changed by these actions, assuming different characteristics and new meanings. In this sense, landscape can be considered an expression of local culture, because its creation is guided by economic mechanisms and socio-cultural values that govern the actions of a certain group of people, and by the technological evolution that makes this process last by giving it a new way to relate to spaces and places of living. At the same time, precisely because landscape reflects on interrelations between population and territory in which it is located, it becomes an element of cultural identity for those who inhabit it (square) and who are part of it.

In the era of the third industrial revolution landscape is perceived not only as a factor influencing well-being and quality of life of its inhabitants, but also as an element of cultural identity, because it nourishes the sense of belonging to a place. At the same time, it strengthens connections between members of a community and contributes to their well-being, sometimes even at an unconscious level. As Viviana Gravano argues, *"what matters is that the perception of what is outside of us, specifically of our context, can no longer be considered neither unique nor unambiguous"* (Gravano V., 2012). Above all, in times dominated by globalization, the standardizing and leveling down effects of which have a major impact on both people's lives, their living environment, and where the connections that link communities to their territory seem to have broken, it is of great importance to educate people, not only superficially, of their landscape.

In fact, accelerated economic development of the last century has had a profound impact on landscape and the environment in general, both on material level, with the establishment of new production systems, modern urban planning, zoning and productive exploitation, and at the perceptive level, forcing an individual to rediscover the sense of belonging and of a shared landscape. There is a clear link between landscape and individual: if anthropic actions transform the environment, this in turn changes the perception of man in regard to it.

In conclusion, when reflecting on landscape as a cultural factor in reference to its anthropization, *"one should never address a generic population, but should define on a case-by-case basis what context one is talking about and what one's point of view is"* (Lai F., 2001), because landscape can be identified as a code according to which one should observe a particular community as a movement of internal and external relations, and as a cultural process that is always in movement, as a site-building and site-based narrative. Landscape observed in its anthropic dimension results to be a space conceived not as a pure container, but as a whole complex of economic, political, social and religious factors that relate to a given environment.

1.5 LANDSCAPE 4.0 NETWORK

DECARBONIZED GRID POWER DISTRIBUTION
- SOLAR POWER
- WATER POWER PLANTS
- BIOMASS PLANTS
- WIND POWER
- GEOTHERMAL
- COAL-OIL-GAS
- NUCLEAR POWER PLANTS

"In the summer of 1969, we were all with the noses in the air to watch the moon ... we believed that our satellite would be launched... to become a copy of our planet. In fact, in those days, without anyone knowing it ... between Los Angeles and Washington, two computers were exchanging information for the first time, and the network was being born, becoming a world within which time and, above all, space ... performed almost no function anymore."

Farinelli F.
L'Unione Europea abolisce l'idea di territorio valida per i secoli. Senza sapere forse perché, ha fatto suo il modello della rete.

IMG. 18
img. by: AMO. Presents Vision for a Decarbonized European Power Grid to EU leaders. April 07, 2010

Farinelli in his article (Farinelli F., 2015) discusses the transition of a modernist concept of territory identified in a space system, where the surface of the whole planet is represented. He provides an objective view of what surrounds us, where landscape became able to be resilient to the changes in progress and has therefore turned central in the contemporary debate, inside a system of interconnected networks becoming an invisible infrastructure that unites us.

Nowadays, landscape assumes a different role compared to the past (as illustrated earlier). Popularized and democratized, today it belongs to everyone, while in the past it took on a role of a social code and a distinctive mark of the privileged class, which in many cases was willingly associated with the sharing of symbolic places or topical representations. From being peripheral landscape has become central (if not indispensable) in philosophy and geography, not forgetting its increasingly important place in sociological, anthropological and archaeological theories.

It becomes the central object in the process of spreading of modifying interventions caused by man, interventions linked to industrialization and economic liberalism, which have given value to the soil, leading to an increase in the risk of landscape deterioration caused by inappropriate interventions to the point of losing its identity.

How, then, should we face the complex transactions happening in our territories and in the cities of the future?

In the European geopolitical framework prevails the tendency of guiding planning and landscape disciplines (both project and theoretical) towards a "post-fossil" context (Sijmons D. et al., 2014). This goal can be achieved through a smart and technologically advanced transformation deriving from the need to detect and utilize appropriate political and projecting strategies in landscape and urban planning. These strategies cover various spheres from energy market to legislation, from adaptation to technological development, and have a transdisciplinary approach that can connect and simultaneously develop technology and urban design.

Therefore, in this transdisciplinary approach we can find answers to our questions, in practice and theory emerging over the last 16 years, during which many personalities of our scientific community have devoted themselves to the exploration of the convincing project called Landscape Urbanism. Landscape Urbanism is transdisciplinary by definition[29], as it brings together the inheritance of landscape planning and the dynamics of contemporary urbanism, integrates knowledge and techniques of environmental engineering, urban strategy, and landscape ecology, while expanding the science of complexity and emergency on digital design tools and ideas of political ecology. Combination of all these techniques and knowledge becomes a new support for project planning within urbanism understood as a social, material and ecological factor, and is constantly modulated by spatial and temporal forces connected into a network.

This recent landscape development goes beyond linguistic and disciplinary boundaries inherent in the territories, and

29.
Landscape urbanism appears to offer a way to consider the complex urban condition; one that is capable of tackling infrastructure, water management, biodiversity, and human activity; and one that asks and examines the implications of the city in the landscape and landscape in the city. James Corner. "Terra Fluxus" in The Landscape Urbanism Reader, ed. Charles Waldheim.

IMG. 19
Landscape and Energy. Designing transition Published on Jun 12 2014. By Sijmons D. et al.

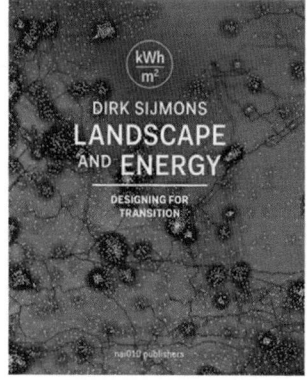

turns landscape into not just a characteristic expression of a uniform world, but above all into one of the essential means that contribute to the globalization of visual concepts and schemes.

If, as states Jakob, "landscape is an artificial product of a culture that perpetually redefines its relationship with nature" (Jakob, 2009), then designers like Corner (for whom the central role lies in the process), or Branzi (for whom the most important aspect of landscape projecting is to renounce to the outline of the project) both seek to respond to the need of projecting something, for what it is not possible to conceive and to see a finished state.

This reinvention also gets transformed through the need to rethink a way of acting on landscape, as suggested by Charles Waldheim.

New horizons open in the world of project and landscape theory. The process of renunciation begins with respect to the fundamental role of figuration in order to focus on the points of connection between the elements, on functioning of the territories, on the processes of growth and modification of landscape.

In this sense, the art history of landscape and the conceptual overturn analysis briefly described in the preceding paragraphs could be considered as fundamental links to the experiences of pure projecting proposed by the currents of Landscape Urbanism (LU).

This discipline finds its origins in Anglo-Saxon culture and in particularly evident interweaving between ecology, landscape and urban design. The LU is situated at the heart of a substantial interdisciplinary research tradition. One of the most well-known theoretical experiences and expressions of the discipline includes authors, lecturers and designers of the level of James Corner and Charles Waldheim , who provide a valid interpretation of the contemporary urban landscape. Mainly as a result of a terminological enrichment, they have proved that it is possible to test various readings and to give better interpretations to the dynamic features of transdisciplinary and multiple proximity of the contemporary urban ecosystem as a place for exchange of energy and knowledge. For the first time, this particular place for exchange and study emerges as a branch of Landscape Ecology and focuses on the organization of human activities within natural landscape (Shane, 2004). In this sense, Landscape Urbanism exhibition (1997)[30][31] by Charles Waldheim seems to be particularly interesting.

In this exhibition, it is illustrated how this discipline focuses on research and deepens connections between human activities and natural landscape: interstitial, infrastructural spaces and ecology are considered as a background of planned and unprogrammed social activities on a public territory.

Waldheim discusses the issues of how urban density emerged from landscape and how ecology and urban technologies could be decisive in defining spaces, where social activity was articulated following the logic of the bottom-up.

30.
Charles Waldheim turned this landscape ecology approach towards the city in a Landscape Urbanism exhibition in 1997, the year he established a new landscape urbanism option for undergraduates at the University of Illinois, Chicago. (Shane D. G., 2004)

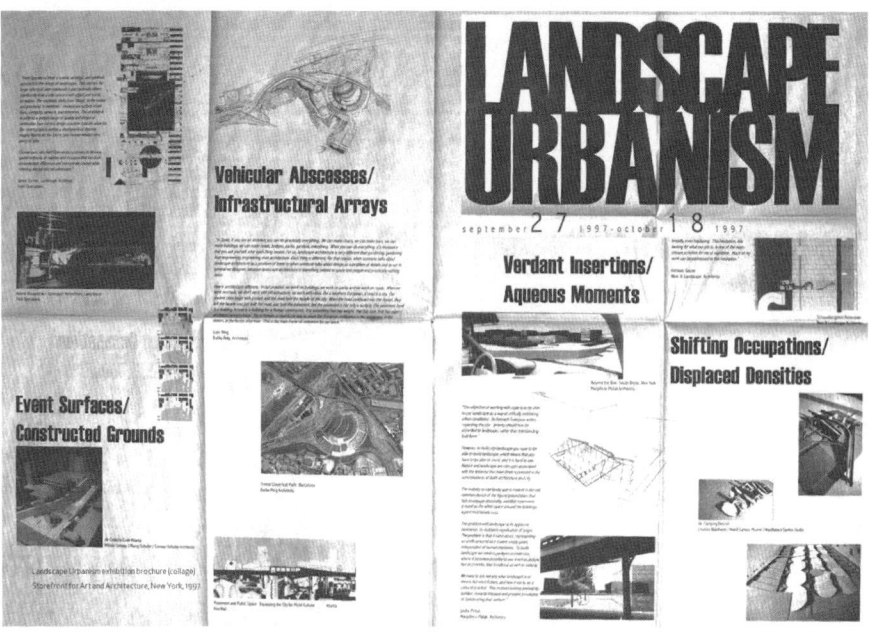

"Lynch's emphasis on large scale thinking has continued in Landscape Ecology and in the Emerging Landscape Urbanism movement, which looks broadly at the organization of industrial society and its use of natural resources as constituting an urban landscape far beyond the scale of the traditionally bounded European city. (...) The Landscape Urbanism movement embodies many of Lynch's global, regional and ecological concerns." (Shane 2005).

31.
IMG. 20
Landscape Urbanism exhibition brochure (collage), storefront for Art and Architecture, New York, (1997). Extracted from the book by Jeannette Sordi: "Beyond Urbanism".

As Grahame Shane points out, current landscape sensitivity is based on awareness of the Dutch and German eco-friendly traditions of planning, which acquired a different perception of the scale of territorial transformations in action with the help of aerial photography.

Landscape Urbanists use to advantage the research of German ecological planners from the 1930s and 1940s with their aerial perspectives, creating a global panoptic vision of industrial and settlement pattern in landscape (...). They also exploited *"The American ecological research into the migration patterns of species"*, which perceived landscape as a series of largely manmade "patches" of agricultural and rural order through which species must flow to migrate. Each "patch" has its own ecology and dynamics, forming a platform of shelter and resources spaced at intervals. There is an obvious analogy with the flow and ghetto of an American immigrant city and the splintering post-modern global patchwork of highly differentiated urban enclaves connected by high-speed communications and transportation routes. (Shane D.G., 2005)

Charles Waldheim in his article "Landscape as Urbanism" supports the need to overcome the obsolete opposition between nature and city several times expressed through the distinct separation between parks and urban fabric. Yet a different point of view was expressed in the urban plot of Manhattan and Olmstead[32][33] Central Park. Instead, the presence of landscape in the city should be considered and facilitated through studying the expansion of a city in its surroundings. James Corner in his "Taking Measures across the American Landscape" realized this phenomenon by verifying the zenithal perspective and finding a highly productive vocation of the American agricultural, industrial and mining landscape, which not only changes landscape in its spatial conception but leads to a sense of transformation. "

Machine city and natural ecological systems have to be put on the same level" (Shane D.G., 2005).

32.
Frederick Law Olmstead, Central Park Plan, 1869.

33.
IMG. 21
Central Park Plan.
By Frederick Law Olmstead, 1869

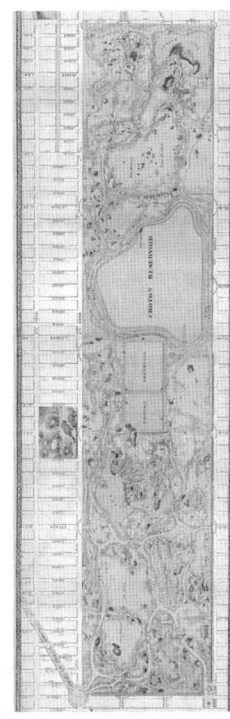

1. FIVE CONCEPTS OF LANDSCAPE

Not to mention Bruce Mau's provocative affirmation: "everywhere is city"[34].

We should rather observe how the extension of urban neighborhoods generates a density that perpetuates itself more and more often, creating homogeneity of a dispersed fabric in, citing Rem Koolhaas, a Junkspace.

"We think Junkspace is an aberration, a temporary solution, but it is a mistake. Junkspace is the reality. It was created in the 20th century, and the next century will become the apotheosis of it." (Koolhaas R., 2006)

Within urban complexity it is possible therefore to recognize both urban density and that of contemporary rural landscapes that could be approached with multidisciplinary ecological logics, which overcome the traditional notions of hierarchy, boundaries and centers, as well as the nature-culture dualism (Mostafavi M. et al., 2003).

LU can thus be defined as an imaginative project that considers the connection between geography and social heritage to be essential inside the study of public relations of collective memory and shared desires. (Church, 2013)

"Collective imagination, informed and stimulated by experiences of the material world, must continue to be the primary motivation of any creative endeavor. ... Public spaces are firstly the containers of collective memory and desire, and secondly they are the places for geographic and social imagination to extend new relationships and sets of possibilities. ...
It seems Landscape Urbanism is the first and last imaginative project, a speculative thickening of the world of possibilities." (Corner J., 2006)

Therefore, we are dealing with the search of a sensitive identity image at different scales of landscape, i.e. an image in progress. Today, we can observe this image in the midst of the fourth digital revolution (4.0) based, as we have already mentioned, on the fusion of new technologies, aimed at fading and disappearing of boundaries between different thematic spheres, from physics to digital biology.

At the level of the scale, we are moving towards a landscape and technological revolution, where landscape is seen as a network

34.
IMG. 22
"Massive Change," 2004, designed by Bruce Mau published by Phaidon.

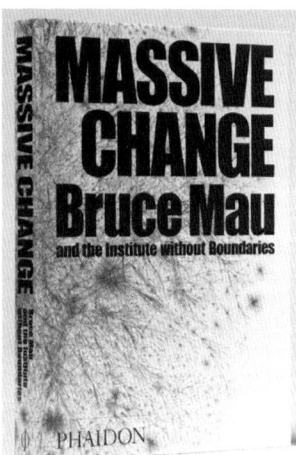

35.
Since the new processes that mobilize and deploy exchange operate through, between and over multiple sites and disciplines – to the point that urbanism, landscape, infrastructure, economics, and information are now inseparable in terms of their influence on the organization of the public realm – they cannot be solely defined through or against traditional design conventions, resulting in a difficulty, therefore, in synthesizing their operation into a new articulation of public site(s). The plasticity of contemporary ecology of exchange has resulted in the connection between public space and commerce progressing from a site/object relationship to a more organizational one that exists "across" or between multiple sites of occupation. It is in acknowledging this shift from "at" (singular) to "across" (plural) site(s), that our uncertainty with the interpretation and territorial articulation of these new processes will be alleviated." (Lyster C., 2006)

of infrastructure-technological, social-cultural, economic, and information exchange networks. The nature of connections (social and energetic), influencing material conformation and specific operations in the area, determines the morphology and the way to occupy landscapes of our urban and suburban contexts (Lyster C., 2006) and those of the countryside.

We are therefore entering (or rather we are already inside) a new process of exchange, through which resources are demobilized and reallocated, operating simultaneously on different disciplines: urban landscapes, infrastructures, energy, economics, and information cannot actually be considered separately when studying or working on organization of public spaces. Such connection and fluidity are difficult to represent with traditional methods of description and planning, but instead it needs a new articulation of material and immaterial spaces.

It is therefore necessary to pass from singularities of theories, actions, interventions, projects etc. to plurality. From this observation derives territorial interpretation of the new exchange processes[35], just as James Corner argues in his concept of *"exchange network system" deepened in the text* "Landscape Urbanism, manual for the machinic landscape" by Mostafavi and Najle (2003).

Every element of the system is in contact with all the others in connection of co-dependence and interactivity within soft systems, which are evolving and sensitive to any change stimulus that allows absorbing, transforming and exchanging information with what is around. The robustness and stability of the stimulus stem from its ability to handle dynamic differential motion processes (Corner J., 2003).

Our vision of landscape is therefore "nourished" by an interest in environmental sustainability, closely related to the application of good European and global practices that guide us towards the goals of sustainability. Though it is not only tied to the balance between produced and spent energy, but also to the harmonious integration in time and various scales of rural and urban landscape of different cognitive and operational networks.

Landscape Urbanism approach aims to develop space-time ecology that encompasses every force and element functionING in the field of action and considers them as a continuous and complex network of interrelations (Corner J., 2003).

As argued by Charles Waldheim, the attention of contemporary research is attracted to the final conceptualization of landscape, which includes an attempt to understand the theoretical relevance according to which sites, territories, ecosystems, networks, and infrastructures are organized in vast urban fields (Waldheim C., 2006).
These elements should be studied with a systemic approach. The complex network of interdependencies, influences, and mechanisms of mutual control should be aimed at creating a moving landscape, which would be considered as a dense network of connections able to activate multiscalar resilient processes.

"Paradoxical and complex, Landscape Urbanism involves understanding the full mix of ingredients that comprise a rich urban ecology."

Mostafavi M., Najle C.
*Landscape Urbanism.
A manual for the machinic landscape.*

Quoting Mostrafavi, it is possible to state that Landscape 4.0 is up to date with today's innovative era. Therefore, it implies an overlay of layers and components that are able to create a rich urban ecology, a steady multi-scale network movement of interventions that binds us to landscape.

IMG. 23
In:Grounding Landscape Urbanism
by Shanti Fjord Levy

Published in Scenario 01: Landscape Urbanism, Fall 2011

Original image published in: Cities in evolution. An introduction to the town planning movement and to the study of civics. by Geddes, Patrick Sir

Published 1915 by Williams in London.

"A NETWORK OF SOCIAL AND ECOLOGICAL CONNECTIONS IS FORECAST TO INTERWEAVE OVER TIME."

Shanti Fjord Levy
Grounding Landscape Urbanism

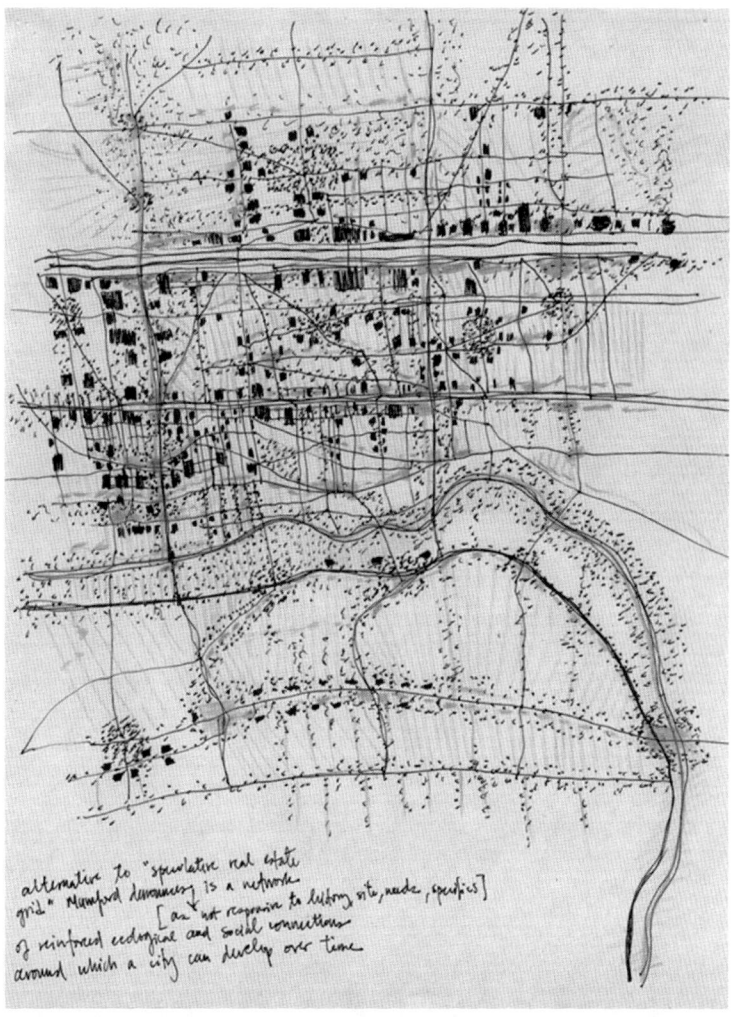

CHAPTER TWO

2

KEY WORDS AND TIMELINE

2.1 KEY WORDS

By selecting some key concepts we intend to outline the framework into which the research is inserted, creating a common ground on which it will be developed.
The goal is not to build a complete and exhaustive glossary, but rather to rebuild an analog of concepts that transmits the meaning of some of them, on which the entire path of research is developed.

LANDSCAPE

In geography, this term indicates the whole of sensitive manifestations of a country or territory, similar to the words **paysage** in French and **landschaft** in German. The latter is often identified with that of "region".
In this research, we would like to take into consideration the complexity of contemporary territories and landscapes by imposing a conceptual and operational leap and entering into the theoretical framework of "Landscape Urbanism". This is done not just by trying to merge the concept of landscape with infrastructures that make part of it but by turning the same landscape into a multidisciplinary infrastructure of the future.

NETWORK

Generally, it defines a set of interconnected entities (objects, people, etc.). It enhances circulation of material or immaterial elements between different and distinct entities according to well-defined rules.
The term "network" is essential for our work, as it consists of nodes and streams of connections that keep cities and landscape linked through infrastructure such as roads, railway lines, smart grids (material), and also energy, communication, Internet, and flying networks (immaterial).
Networks can be linked, studied and correlated in different ways. They assist in analyzing future interests of a city and landscape in order to overcome a modernistic planning approach.

SMART

> *"To be smart means: cooperate rather than compete, create a system rather than dominate, put in continuous relation. Does digital technology not state that being same is being connected? I could also use another formula: managing profits, socializing processes, or better managing processes to socialize profits.*
> *To accompany processes means that smart logics tend to be non-taxing, to not predetermine socio-economic and political phenomena, to engage in "listening" logic, in short, smart logics introduce new forms of democracy."*
> *(Bonomi A., Maserio R., 2014, p. 126)*

Objects or dynamics could be considered smart when they respond quickly to various situations and are capable of changing according to the context. Today, the world experiences continuous and rapid evolution, and the key to becoming and remaining competitive resides in being quick in adapting to changes and new realities, in short, "in being up to date".

SMART CITY

> *"a city can be defined as 'smart' when investments in human and social capital and traditional (transport) and modern (ICT) communication infrastructure fuel sustainable economic development and a high quality of life, with a wise management of natural resources, through participatory action and engagement".*
> *(Caragliu, Del Bo e Nijkamp, 2009, p. 7)*

SMART GRID

> *"Smart grids represent the power grid model of the future, which will allow better management of energy streams. Not only electrical current will flow, but also information. The new infrastructure will allow the user not only to receive energy services, but also to send data, which can be shared with other users, to the network".*
> *(Ilaria Bertini, Distributed Generation Manager of the ENEA Research Center, explains the intelligent networks)*

Infrastructures that incorporate "smart" elements in different ways can also be considered as a system within which energy can reverse its flow. This happens when intermittent energy sources of non-programmable origin, for example, wind or solar, are inserted into a network. As mentioned above, the most important element for the user is the possibility to send data automatically to the operator with the help of devices such as smart meters. This information can be transferred to the Service Manager but also shared with other users.

With the same functioning of communication and management of the information, in the present work the management and communication system of the "smart grid" could be used as a system of landscaping and urban planning.

ECOSYSTEM

It represents a union between living organisms and abiotic components of an environment, which interact with the outside world, where energetic and nourishing fluids connect various elements, through them defining what we call ecosystem.

Considering landscape as a heterogeneous space composed of different geographies and ecosystem patterns, by studying the ecology of landscape we can study iterations between different elements and hypothesize possibilities for improvement. In our case, ecosystems guide us through a reading of the context capable of providing the necessary support for the construction of an energetically-efficient landscape, which at the same time is able to unite energy, landscaping and social ecosystems.

INSTRUMENT

Within the disciplines of architecture, urbanism and landscape, diverse tools are used to create a project, and the more technology and anthropology develop, the more various tools evolve allowing creation and elaboration of objects unthinkable in past. We could consider that tools, as already mentioned above, are able to generate devices, which from a philosophical point of view can be considered as a "skein, a multi-linear set, composed of lines of different nature... subject to variations" (Deliuze G., 2007).

DEVICE

It is a term that can have different definitions depending on the context in which it is used. Generally, it is associated with devices or instruments that have certain specific functions.

In our case, the devices of our interest are those that are identified with the acronym ICTs (Information and Communication Technology). Today, they have become a sort of fundamental capital that guarantees cities and landscapes their strong competitiveness. In a context in which the concept of a "smart city" is considered as an indispensable phenomenon of our era these devices become strategic triggering and understanding modern factors of urban production in a common and avant-garde framework.

CATALYST

In Chemistry, a catalyst is a mediator and promoter, able to favor a chemical reaction while remaining unchanged, as opposed to the reagents. Its use becomes of considerable importance in some chemical reactions, which would hardly happen in its absence, or would employ days or years to complete. Therefore, we can affirm that its role is to reduce the time of reaction of a certain action and to favor its fulfillment.

We should consider the fact that today's rural and anthropized landscapes are no longer a target, but a starting point. Spaces open to cities and territories, not defined, without a future destination, or crossed by roads and energetic infrastructures can become a new key to reading cities and landscapes of the future thanks to applying devices or projects that become real urban catalysts in these places. These catalysts trigger actions or provide instruments to trigger new processes in the development of cities and landscapes.

DRIVER

In the economic field, drivers are determinants of the value creations (Value Drivers) that enclose the whole of the factors and variables. They are able to act, influence and stimulate the capacity of various business units, into which a company is divided, with the goal of increasing its global value. These drivers can be classified in two groups: qualitative factors, or strategic value drivers connected to the strategic context of the company, and quantitative factors, or financial value drivers,

which refer to the analysis of accounting documents, use mathematical/estimation formulas and relate to the financial sphere of the company.
In a context of urban and landscape reactivation, drivers can be considered qualitative or quantitative depending on the actions that are to be developed in a certain context. These actions stimulate and influence the capabilities that devices or concepts intend to apply to a given context.

INFRASTRUCTURE
Literally, it would be a structure or complex of elements that constitute the base of support (or an underlying part) of other structures. It is also applicable to society in a figurative sense. That is, infrastructure is a crucial element of landscapes that we inhabit. However, we must not forget that infrastructure often limits connections between natural and artificial elements, which respond only to anthropic necessities that dictate the pace of our lives and our relationships. The connections that are created to compensate for a "need" and the thoughts that arise behind the creation of infrastructure must change radically as a result of exploring the paradigms of contemporaneity. We must think of infrastructure as a hybrid, high-functioning and multi-functional complex in strong relation to landscapes that it crosses

2.2 TIME TABLE

IMG. 24
Time Table by Giulia Garbarini

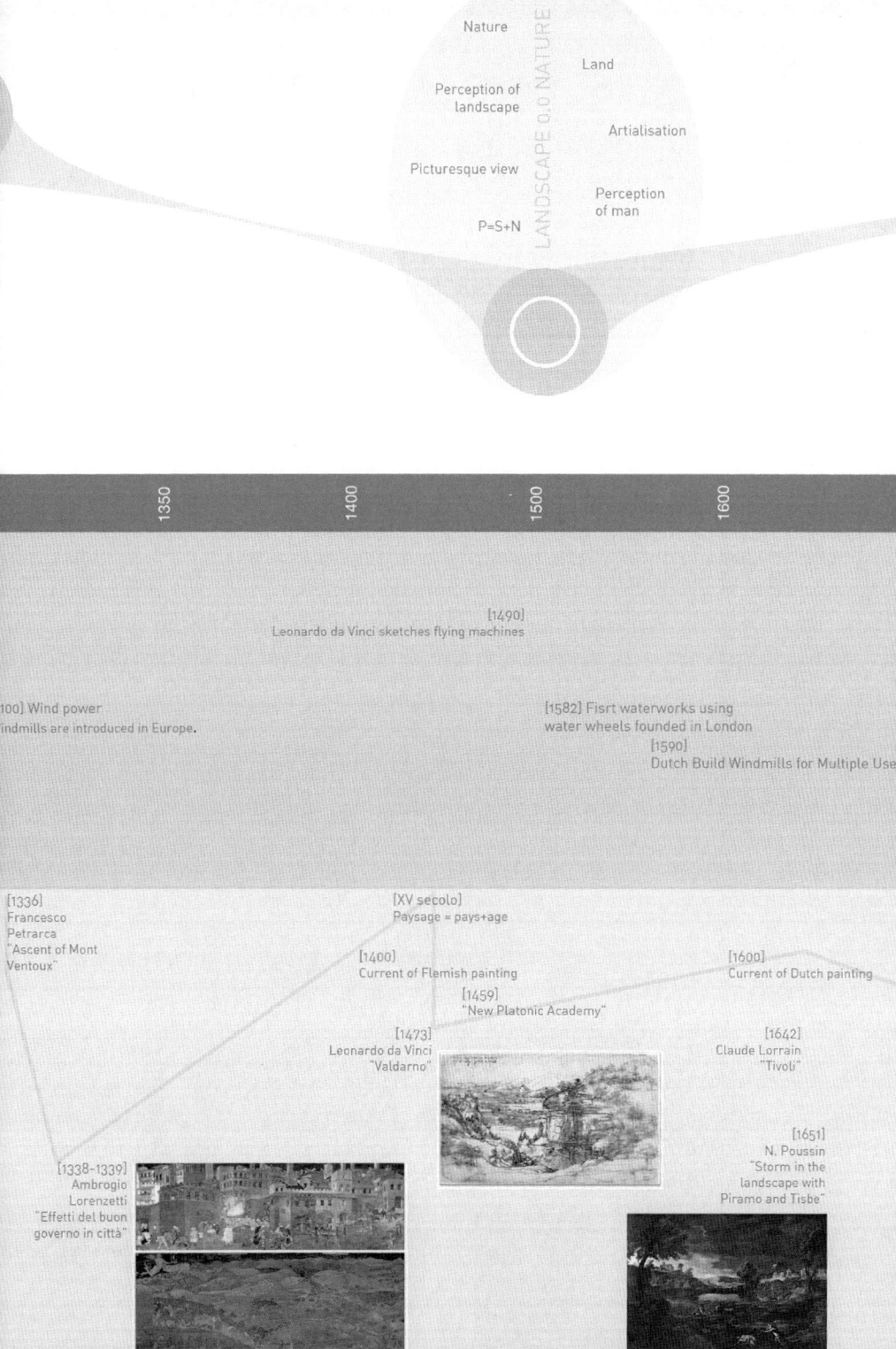

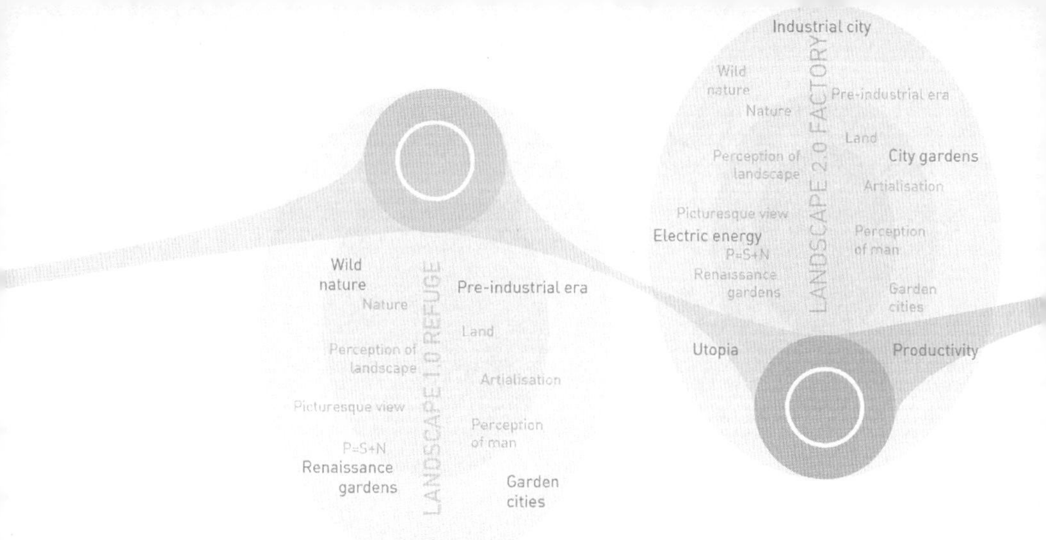

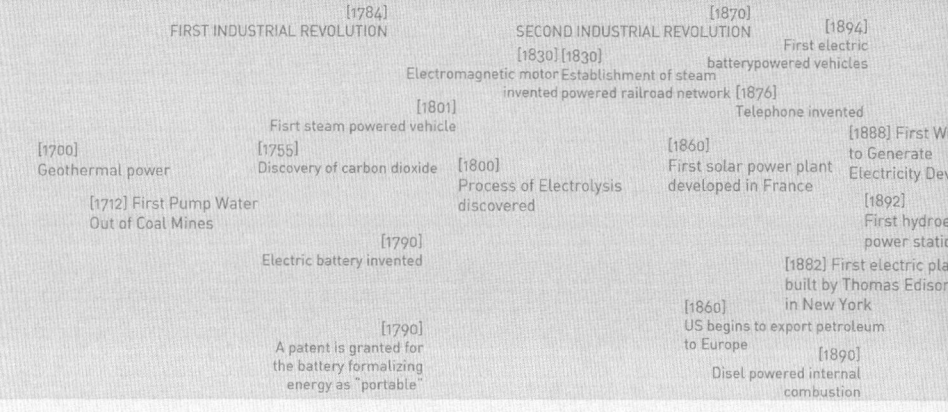

Industrial city
Wild nature
Nature
Pre-industrial era
Perception of landscape
Land
City gardens
Picturesque view
Artialisation
Electric energy
Perception of man
P=5+N
Renaissance gardens
Garden cities
Utopia
Productivity

LANDSCAPE 2.0 FACTORY

Wild nature
Nature
Pre-industrial era
Perception of landscape
Land
Picturesque view
Artialisation
P=5+N
Perception of man
Renaissance gardens
Garden cities

LANDSCAPE 1.0 REFUGE

| 1700 | 1750 | 1800 | 1850 | 1900 |

[1784] FIRST INDUSTRIAL REVOLUTION
[1870] SECOND INDUSTRIAL REVOLUTION
[1894] First electric batterypowered vehicles
[1830] Electromagnetic motor invented
[1830] Establishment of steam powered railroad network
[1876] Telephone invented
[1801] First steam powered vehicle
[1700] Geothermal power
[1755] Discovery of carbon dioxide
[1860] First solar power plant developed in France
[1888] First Wi to Generate Electricity Dev
[1800] Process of Electrolysis discovered
[1712] First Pump Water Out of Coal Mines
[1892] First hydroe power statio
[1790] Electric battery invented
[1882] First electric pla built by Thomas Edison in New York
[1860] US begins to export petroleum to Europe
[1790] A patent is granted for the battery formalizing energy as "portable"
[1890] Disel powered internal combustion

[1700-1800] Current of English painting
[1800] Current of French painting
[1898-1902] Howard Garden cities of To

[1651] Treatise on Painting

[1726 - 1727] Canaletto "The Reception of the French Ambassador in Venice"

[1873] C.Monet "Poppies"

[1818] Caspar David Friedrich "The Wanderer above the mists"

[1742 - 1745] Canaletto "View of the arch of Constantine and Colosseum"

[1909] P. Picasso "Factory in Horta De Ebro"

[1888] V. Van Gogh "Sower at sunset"

SMART LANDSCAPE

LANDSCAPE 4.0 NETWORK

LANDSCAPE 3.0 SQUARE

Diagram keywords (left cluster):
Land, Nature, Wild nature, Industrial city, Council of Europe, Perception of landscape, Place, Picturesque view, Landscape Concept, Electric energy, Double dimension of landscape, Renaissance gardens, P=S+N, Garden cities, Pre-industrial era, City gardens, Artialisation, Perception of man, Productivity, Space, Utopia

Diagram keywords (middle cluster):
Land, Nature, Wild nature, Smart, Perception of landscape, Pre-industrial era, Electric energy, Industrial city, Council of Europe, City gardens, Artialisation, Picturesque view, Perception of man, Post-fossil, Double dimension of landscape, Productivity, Space, Ecology, Renaissance gardens, Utopia, P=S+N, Place, Landscape Concept, Garden cities

Diagram keywords (right cluster):
Landscape Urbanism, Strategy of surface

Timeline

1950 — 2000 — 2050 — 2100

[1905] Photoelectricity invented
[1969] THIRD INDUSTRIAL REVOLUTION
[TODAY] FOURTH INDUSTRIAL REVOLUTION

[1905] Albert Einstein First theory on photoelectric effect
[1951] UNIVAC computer developed
[1981-85] First scace shuttle lauch / Introduction of personal PC and launched of mobile phone

[1956] Fiber optics
[1956] Photovoltaic cell developed
[1993] First wind turbine
[2050] Energy strategy milestone

[1939] Nuclear power
[1942] First nuclear fission reactor
[1956] First geothermal power plant
[2007] IPCC Report
[2014] Rockefellers and over 800 Global Investors Announce Fossil Fuel Divestment

[1913] Harry Ford intriduces the concept of industrialization
[1920-40] Construction of hydroelectric plants in the American Southwest
[1965] Introduction of micro computer

[2000] European Landscape Convention / Corner J. Taking Measures Across the American Landscape

[1932] F.L. Wright Broadacre City
[1969] Archizoom No-Stop City
[1994] Branzi A. Acronica
[2006] Waldheim C. The landscape urbanism reader

[1965] Peter Cook Plug-In City
[1983] Koolhaas R. - Villette Park
[2008] Eco-city - MVRDV

[1923] Paul Klee "Landscape with yellow birds"

[1983] Cristo Javacheff "Deladelmur"

[1962] A. Warhol "Do it yourself"

[1984-1989] A. Burri "Cretto di Burri"

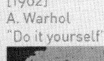

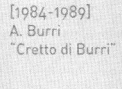

2. KEY WORDS AND TIMELINE

CHAPTER THREE

SMART LANDSCAPE

The guiding principle of this research project lies in a combination of two different disciplines. That is, correlation between two vast disciplinary areas: landscape with its conceptual evolution, on the one hand, and, on the other, the term "smartness", overused and almost omnipresent in various cultural and social spheres. The purpose of this research is to illustrate how cities, peripheries and territories are currently integrated into a single network of interconnections that will create "Smart Landscape".

Landscape, which has entered the new millennium with a new interpretation, with a new "mission" and function, is no longer just a field of experimentation for visual arts, a beautiful garden, an urban green area within a city. Landscape, from the early 2000s to present, has been seen as "everything"[36], and, unlike geographical "place", it is not composed of things but is only a way of seeing and representing objects in the world (Farinelli F., 2003). In fact, the power of this discipline resides in its generic nature; it is strong to a point that it leads to being "matched" with a multitude of disciplines that approach and enter landscape, redefining it successfully, due to its generic feature, which in turn becomes its potential. Declining various contents of other disciplines within landscape, the latter occasionally acquires its own specificity, transforming and projecting new landscapes accurately in relation to the characteristics and needs of the site under consideration and no longer imposing conventional and standardized practices from above.

Here, we have "smartness" consisting of technologies, but also of new and unpublished social and cultural models. It seems difficult to identify (at least in the optics of this research) a unique and somewhat shared definition of "smart" and even less that of "smart city", "smart land" and "Smart Landscape". The intelligence that resides in landscapes, however, can not only be tied to a technological factor or "big data" when talking about its ability to leverage various devices and tools and link them into a vision that functions by combining points and successful interventions, as was already anticipated in the No-Stop City of the Archizoom.

By using smart grid architecture as a network on which to construct smart landscape we can exploit its rigidity as a potential. Thus, network concept is to be seen as a contempo-

36.
"What was worth (and still is) for a city just a century ago, according to Adolf Loos'"Ins Leere Gesprochen", today is worth for landscape [...] Landscape is an unlimited issue, its existence poses the problem of how it can be considered a whole that is at the same time visible but without boundaries and therefore not measurable, and precisely for this reason it implies an extreme difficulty: the question of totality. In the meantime, it is worthwhile distinguishing landscape from all the other models (territory, space) existing on the Earth, being of delimited nature, with which it has recently been effortlessly and hastily identified."(Farinelli F., La capriola del paesaggio. 2015)

rary urban condition rather than a concrete planning method, where everybody seeks to redefine their perception of cities and urban areas in the age of globalization, society of networks and virtual technologies.

To come to the definition of the "Smart Landscape" concept we would like to try, in advance, to answer some of the questions that lead us to the perception of this concept:
Why does this research work fit into the "smart" context? What do we mean by "smart"?
And yet, the questions that follow the arguments of this work are: why are we not talking about a "smart" city then? Or "smart" land? Why are we talking only about "Smart Landscape"?

3.1 SMART

Today, the word "smart" has become one of the most "trendy": it can be encountered in our everyday life, in administrations and in the world of work. As often happens, it is a term that originates from the English language, and it has been overused in many different ways around the world. It is an adjective that has numerous synonyms, such as: quick, skillful, sharp, brilliant, awake, intelligent, etc.

When "smart" is used in reference to a person, it describes their intelligence together with their abilities or learning capacities and response rate to external stimuli. But it could also be used when talking about objects or dynamics, which react or respond to various situations very quickly and are capable of changing according to the context. In a perspective of improvement, they are considered "smart". Today, the world experiences continuous and rapid evolution, and the key to becoming and remaining competitive resides in being quick in adapting to changes and new realities, in short, "in being up to date".

To be smart does not only mean to be intelligent but also ready, awake, brilliant, responsive: in other words, it means to be able to show adaptation skills, abilities of problem solving and quick learning in various situations. This adjective is often used in combinations, such as "smartphones", literally "intelligent phones", i.e. phones with advanced features: they become devices that incorporate functions of a hand-size computer and a cellular phone. More and more often, such words as smart watch, smart TV, smart work, and, moreover, smart city, smart economy and even smart governance are used, where "smart" contains the concepts of better quality of life and less environmental impact as a result of intelligent use of technologies.

To understand the philosophy of "smart", it would be relevant to study the philosophy of "goals". Goal setting, like other behavioral philosophies, dates back to the Greeks, to Aristotle and Plato, who's philosophies on final causality suggest that "purpose can incite action" (George C. S., 1972). However, through the centuries, human beings have rarely written and documented management and organization techniques. The idea of creating smart goals was never formalized until much later. Scholars and practitioners claim Peter Drucker's book "The Practice of Management" (1954) is fundamental in launching the development of the acronym smart (Morrison M., 2010). However, Drucker did not make any direct reference to it in his book (Morrison M., 2010). During the 1940s and 1950s, various engineering and educational publications were discussing the merits of specific and measurable goals (Morrison M., 2010). In early managerial and educational publications, organizational activities were labeled with specific and measurable characteristics, which suggest that smart goals and their acronyms were already known and fairly prevalent. The use of a variety of words to describe goals by both educational and business worlds suggests that the "smart" acronym emerged organically rather than having been specifically invented (Morrison M., 2010).

SMART goals are written using the following guidelines: 1) Specific – define exactly what is being pursued; 2) Measurable – there should be a number to track completion; 3) Attainable - the goal should be achieved; 4) Realistic – feasible from a business perspective; 5) Timely – they should be completed in a reasonable amount of time (Williams, 2012). Over time, the "smart" acronym experienced changes as more and more people were noting the benefits of the concept.

IMG. 25
Grottaglie (Taranto),
Murales By writers BLU.

"To be smart means: cooperate rather than compete, create a system rather than dominate, put in continuous relation. Does digital technology not state that being same is being connected? I could also use another formula: managing profits, socializing processes, or better managing processes to socialize profits.
To accompany processes means that smart logics tend to be non-taxing, to not predetermine socioeconomic and political phenomena, to engage in "listening" logic, in short, smart logics introduce new forms of democracy."
(Bonomi A., Maserio R., 2014, p. 126)[37]

Thus, subsequent definitions and use of "smart" were created. For example, Hersey and Blanchard (1988) used the term "smart goals" in the 1988 version of "Management of Organizational Behavior". Others claim that George T. Doran (1981) developed the concept of smart goals in the discipline of project and program management.

Regardless of from where the concept of "smart goals" originated, businesses, administrations and universities have found it to be a valuable tool. Various training companies have developed models on how to effectively apply it. An Internet search of these companies can provide insight into the application of smart goals. For example, the website of "Time Management Success" (2017) gives tips for managing time to become more effective, and similar information could be found on the website of Wayne State University (2017).

In this doctoral research context, the "smart" term aligns more with the statement made by Aldo Bonomi in his book "From Smart City to Smart Land" (2014). At the same time, we only find inspiration in it not fully adhering to his concept of smart land.

What we accept in Bonomi's proposal is the definition of "smart" referring to the mode of operating, mostly in such disciplines as architecture, urbanism and landscape. "Smart" operating helps create a system, a network of connections, shared actions and processes that do not dominate over or interrupt each other. On the contrary, they are balanced and united into a system by constantly establishing a series of profitable connections that activate fast, dynamic and intelligent processes.

The "smart" logic and dynamics that could be found in landscapes should generally set a new idea of citizenship, city, and landscape as the central axis of their ideation and dissemination. Such logic should be based on connections different from any of those that appeared during the Modernist era, between public and private sectors but also between technology and context, by allowing and trying to facilitate smart community dynamics.

Considering that smart logics are generated by and for technology, they can be, like all devices and tools, activated and planned in the name of a particular need to fulfill or instrumentation to activate, but at the same time, they may be of "diffusive" nature (Bonomi A., 2014, p. 126) and potentially

37.
"Essere smart vuol dire: collaborare anzi che competere, fare sistema anziché dominare, mettere continuamente in relazione. La così detta tecnologia digitale non afferma che l'essere stesso è relazione? Potrei usare anche un'altra formula: governare i profitti, socializzando i processi, o meglio governare i processi per socializzare i profitti. Accompagnare i processi significa che le logiche smart tendono a non essere impositive, a non predeterminare i fenomeni socioeconomici e politici, a innestare logiche "dell'ascolto", insomma le logiche smart aprono a uove forme di democrazia." (Bonomi A. Maserio R. 2014 p. 126)

transparent. This way, such technologies or systems (that we can define as enablers) are integrated into social, urban and landscape systems, reinforcing and disintegrating them at the same time.

According to Arthur C. Nelson (2002), "smart growth" is about conserving open space, compacting mixed-use development, revitalizing old centers, enhancing public transport networks, and sharing development costs equitably. These principles are also useful to new urbanism, although it should be considered that new urbanism covers the formal aesthetics of development, while smart growth is more about the socio-economic and political processes that give rise to extensions in the first place. That is to say, smart growth maintains its place in planning and policy in controlling the city.

Landscape Urbanism has far less power in controlling urban dynamics, but LU and smart growth share their interest in engaging processes rather than superimposing forms. As Corner explains,

> *"the promise of landscape urbanism is the development of a space-time ecology that treats all forces and agents working in the urban field and considers them as continuous networks of interconnections" (Corner J., 2006, p.30).*

Nelson states something similar, describing smart growth as *"a systems approach to environmental planning, shifting from development orientation to basics or ecosystems planning"* (2002, pp.88–89).

The above-mentioned statements allow us, in such a sense, to dematerialize the single decision-making power and propulsor of change favoring a new one. This "power" is no longer linked to property but to the possibility of sharing, a sort of a "rent" (such as bike and car sharing), which, within their mechanism, have those Smart exceptions and features that components of a Smart landscape should have.

3.2 CITY

IMG. 26
City of Leonia.
Imm. by: Alessandro Armando, Francesca Ballarini and Elena Nuozzi.

"The city of Leonia refashions itself every day: every morning the people wake between fresh sheets, wash with just-unwrapped cakes of soap, wear brand-new clothing, take from the latest model refrigerator still unopened tins, listening to the last-minute jingles from the most up-to-date radio. On the sidewalks, encased in spotless plastic bags, the remains of yesterday's Leonia await the garbage truck. ...you can measure Leonia's opulence, but rather by the things that each day are thrown out to make room for the new. So you begin to wonder if Leonia's true passion is really, as they say, the enjoyment of new and different things, and not, instead, the joy of expelling, discarding... Nobody wonders where, each day, they carry their load of refuse. Outside the city, surely; but each year the city expands, and the street cleaners have to fall father back. The bulk of the outflow increases and the piles rise higher, become stratified, extend over a wider perimeter. ... Leonia's rubbish little by little would invade the world, if, from beyond the final crest of its boundless rubbish heap, the street cleaners of other cities were not pressing, also pushing mountains of refuse in front of themselves. Perhaps the whole world, beyond Leonia's boundaries, is covered by craters of rubbish, each surrounding a metropolis in constant eruption."

*Italo Calvino, Invisible city, 1993.
Pp .111 – 112*

Why not to talk about a "smart city"? Before answering this question, we must ask ourselves what a "smart city" is.

In response to climate change, increase of atmosphere pollutants and constant consumption of soil that occurs due to expansion of cities (with the awareness that the latter is the direct reason for the above-mentioned issue, according to Italo Calvino), the European Union has been encouraging member countries to develop and adopt tools and/or policies that have the potential to help create smart communities, which thanks to integrated and sustainable solutions could be able to offer clean and safe urban and energy environments.

As highlighted by the "Smart Cities Atlas" (AA. VV., 2015), the European Union's idea of the development of smart cities can be found in the Strategic Energy Technology Plan, more commonly known as SET-Plan, which envisages construction of thirty smart cities in Europe by 2020.

The concept of "smart city" was created around the fourteenth century with the birth of Renaissance cities, extremely similar to contemporary smart cities. Both now and then, these cities were developed with the same intent: to completely revolutionize architectural and urban planning.

In the contemporary era, creation of the theoretical basis of a smart city concept coincided with the birth of cybernetics developed by the mathematician Norbert Weiner of the Massachusetts Institute of Technology[38] (Van Timmeren et al., 2015).

In fact, for Hall (2000), a smart city is a city that *"controls and complements the conditions of all its critical infrastructures including roads, bridges, tunnels, rail/subways, airports, ports, communications, water, energy, buildings to better optimize the resources, plan maintenance activities, and monitor security aspects, improving services provided to citizens".*

For Aldo Bonomi et al. (2014), who also discuss the concept of smart land,

"a smart city is a city of the future, where fewer resources produce more services for citizens and businesses, using the most advanced technologies and management systems, reduce waste and negative impacts, be they environmental, economic or social. ... A smart city is an organic city, a system of systems, which in the urban space faces the challenge

---

[38]
Cybernetics can be understood as an interdisciplinary field that uses detection, monitoring and the related feedback to model systems and their structures in order to organize and control them efficiently. Within the cybernetic model, all the machines can be interpreted as a balanced network of data flows the components of which can be represented by a set of equations and transformed into a computer simulation that reproduces the behavior of a complex system. After inserting the data into a computer, an analyst can use this model of reality to make predictions about the system by modifying the inputs and observing their impacts.

of globalization in terms of increased competitiveness, attractiveness ..."

Therefore, we can agree with the concept of a smart city given by Caragliu, Del Bo and Nijkamp (2009, p. 7):
"a city can be defined as "smart" when investments in human and social resources, traditional (transport) and modern (ICT) communication infrastructure fuel sustainable economic development and a high quality of life, with a wise management of natural resources, through participatory action and engagement".

Agreeing with what is said previously and investigating the topic of a smart city, we find the conceptualization of three key factors of an intelligent city in the statements of Theresa A. Pardo and Taewoo Nam:
"technology (hardware and software infrastructures), people (creativity, diversity and education) and institutions (governance and politics). Considering connection between the factors, a city is seen as intelligent one when investments in human/social capital and IT infrastructures fuel sustainable growth and improve the quality of life through participatory governance" (Nam T., Pardo T. A., 2011)[39].

Obviously, definitions and conceptualizations of a smart city are numerous, and in this research we have only mentioned those of greater interest. One of them is the Smart City model suggested by the European Union (www.smart-cities.eu). It consists of several categories: "Smart Economy, Smart Mobility, Smart Environment, Smart People, Smart Living, and Smart Governance". The concept of "Smart Environment" is particularly interesting for our research because it is broader than the concept of a smart city. It can be defined, according to Cook (2005), as *"a small world where all kinds of smart devices are continuously working to make the life of the inhabitants more comfortable".* In this sense, "smart environment" can be defined as an environment able to acquire and apply information on its own conditions and be adapted to its inhabitants in order to improve their experience in this environment. In fact, in his book "Reimagining Urbanism" (2014), Maurizio

[39]. ..."technology (infrastructures of hardware and software), people (creativity, diversity, and education), and institution (governance and policy). Given the connection between the factors, a city is smart when investments in human/social capital and IT infrastructure fuel sustainable growth and enhance a quality of life, through participatory governance" (Nam e Pardo, 2011).

Carta argues that introduction of smart cities has completely changed the way of working on urban contexts. Today, it means working only with a "technological prostheses" of a city that aims at reaching the nearest future while keeping up to date. It also means dealing with the European smart city protocols, creating, adapting and improving tools and actions that only work with the technological dimension of landscape but not with environmental and social ones.

In this way, due to the widespread and extensive use of advanced technologies, urban territories are able to deal with a series of problems and needs in a new and integrated way. But how can this way of working be correct if only the technological innovation is taken into consideration, but not the social one?

It would be necessary, as Carta (2014) supports, to work on the technological dimension of participation and on 'open source' information sensors, which form a deliberately incomplete system, where citizens themselves exploit it to provide useful information for its application. This way, certain devices can be employed to improve these contexts and to create smart environment.

When we reason in the terms that derive from the concept of smart cities, we are actually working in the context of a smart community. Therefore, through mingling territory, tangible and intangible infrastructures and humans we are moving towards creation of a univocal meta-organism as a sum of individual parts from which to learn and on which to work. This meta-organism can be nothing but landscape, since landscape itself (according to the recent theoretical research) is a transdisciplinary phenomenon, which represents the union of concepts of nature and society.

IMG. 27
In Sunmagazine.it
Added in August, 5,
by Federica Frezza,
under: Renewable
Energies,Green
Business, Smart City

3.3 LAND

This paragraph focuses on the term "Land", and one of its scopes is to clarify what it expresses/encloses and why in the present work it is not intended to investigate the theory of "Smart Land" but rather "Smart Landscape".

Therefore, it is necessary to define the differences that exist between the terms "Environment", "Landscape" and "Territory". As Rosario Assunto (1973) explains in his book "Landscape and aesthetics", these three terms are often replaced (especially in Italy, where the legislative differences are very weak), as if they were expressing one and the same concepts. This is to be considered as a great mistake as it makes academic discussions absolutely sterile and reduce them to a merely literary discourse.

ENVIRONMENT

1. a place, a physical space, a biological conditions in which an organism (man, animal, plant) lives: natural, artificial environment; marine environment, environmental protection, set of laws and measures aimed at protecting the natural environment of a territory (air, land, water, natural beauties) from any kind of pollution or modification.

LANDSCAPE

1. appearance of a place, of a territory as it seems when observed: a picturesque, animated landscape; admire the landscape; enjoy the beauty of the landscape;

TERRITORY

1. part of land of considerable extension; geographical area: a mountainous, flat territory; in ecology, environment: the problems of the territory;
2. area that constitutes a jurisdictional and administrative unit: national, regional, municipal territory;

www.garzantinilinguistica.it
Accessed 20 June 2017

According to these definitions, the three concepts are interconnected. "Environment" has two main meanings: one of them refers to a biological aspect that depends on various internal conditions, on latitudes and longitudes, and one – to its historic and cultural aspects that make prevail cities or countryside, factories or agriculture. Environment can also be seen from an ecological point of view: it is defined by biotic and abiotic ecological factors that have a direct and significant influence on an organism or organisms of reference; from an economic point of view: it is a place of intersection of vertical and horizontal connections that promote local development; from a point of view of urbanism: it can be viewed as an environmental context, as a set of conditions that in a certain way structure the everyday practices. In the past, environment constituted a fixed point in the projects of physical transformation of cities and territory. Planning often put more emphasis on morphological invariants, status of places, compatibility, performance requirements, suitability, physical networks, geographical references, and spatial information, than on an overall vision, which would lead to development interventions and successful protection of a place.

Another ambiguous concept that has been developed over time is the concept of landscape. It can be defined as a cultural elaboration of a specific natural environment (Sereno, 1983), in which there exists a misunderstanding between the subjectivity of an observer, outsider, often decision-maker, and of those who live and construct the landscape, insiders, those, who accept the decisions. We can denote that landscapes respond to the needs of identification and recognition, spatial anchoring, rooting and collective values. The idea of landscape responds to the ultimate reasons of inhabiting a place, and represents *"an essential aspect of life situation of the population, which contributes to elaboration of local cultures and which represents a fundamental component of Europe's cultural and natural heritage"* (Council of Europe, 2000, CEP).

After having explored the concept and topic of landscape, we finally get to the idea of territory, which in this case is expressed by the word "Land". It has mainly spatial meaning and represents a more or less large area of terrestrial surface that is delimited by geophysical borders, linguistic differences, and administrative policies. As they inhabit their territories, human communities produce values that at least partially evolve from environmental data, because they derive from the continuous mixing of social and environmental processes. However, at the same time, "human" territory evokes different forms of social communication: the jurisdiction of a city; the religious unity of the ancient city; city providing goods and services to its surroundings. For many years, there has existed a prevalence of territory when it came to design, but with the loss of perception of borders, urban diffusion, rururbanization, periurbanization, suburbanization, metropolitan sprawl, and unbound metropolis it is still common to consider territory to be the only field of action.

As we can see, and as the authors of the book "From smart city to smart land" make us understand, the terms "environment", "landscape" and "territory" are linked in a circle, forced to chase each other without ever finding the right definition. They are overused and deprived in the name of their protection. We could simply say that territory is created according to administrative rules and infrastructures, environment – to cynics of the cosmogonic and artificial elements, and landscape – to the culture of man (Bonomi A., Masiero R., 2014, p.123). This "simple" distinction is relevant in the discussion, because it does not let us confuse or incorrectly accept the definition of "smart land" which should not be seen as an extension of a "smart city".

The proposal of Aldo Bonomi and Roberto Masiaro has a political value that stimulates the search of homogeneous territorial configurations or ones to be unified on lower levels in such a way as to recompose them into a smart territorial planning. Born as an adaptation of the logic of a smart city, it is extended and articulated to overcome the traditional division between city and countryside, to introduce the digital mode of production and to allow a smart recomposition of an intermediate company. This process creates links between citizens and governments that act (for and on) the territory in a smart way.

According to this vision, the authors want to know what the most appropriate political forms to achieve the smart objectives are.

As previously mentioned, their proposal is:

> "a smart land is a territorial area in which, with the help of widespread and shared policies, competitiveness and attractiveness of the territory is increased with a particular focus on social cohesion, diffusion of knowledge, creative growth, accessibility and freedom of movement, usability of the environment (natural, historical-architectural, urban and diffused), and quality of landscape and life of citizens"
>
> (Bonomi A., Masiero R., 2014).

Territory has a strong administrative value and is linked to infrastructures. Nonetheless, despite agreeing with the authors on the need to promote smart trends both with respect to urban, territorial, administrative and management logics, it is landscape (and not just territory) that becomes one of the main topics of today's discussions, where different territorial communities are connected into a network and merged into a new configuration of local powers.

Landscape has the power to incorporate all five principles of a Smart City[40]. Having acquired a higher value thanks to the European Landscape Convention, it emerges as a strategic element on which to focus on in smart development. For this reason, in the present research work, we would like to go beyond territory and combine landscape with smart evolution.

40.
Mobility, Environment, Tourism, Culture and Economy. Source: European Program, Smart Cities.

IMG. 28
Less is more
Manifesto for a "smart" society
by Federico Della Puppa and Roberto Masiero

In: Bonomi A. Maserio R. Dalla smart city, alla smart land 2014

3.4 NETWORK

> *"As an adjective, the term 'network' is quite successful in negotiating the multi-dimensionality of the city... The innumerable parts of the city... are recognized and then connected to a larger whole – the network. Such a rhetorical move has theoretical power, making it one of the most attractive metaphorical constructions currently present in the urban theory, if not (yet) a new paradigm."*
>
> *Beauregard (www.metropolitanstudies.de/ leadmin/ lestorage/Network_City_Series_Stand_02-20-07.pdf)*

It is possible to encounter a theoretical view on networks (cities) in works by a range of scholars such as Stephen Graham, Saskia Sassen, and Ann Markusen. A network is to be seen as a concept of contemporary urban condition rather than a concrete planning method, where everybody seeks to redefine *"their perception of cities and urban districts in the age of globalization, society of networks and virtual technologies"* [41].

This position thus represents a greater discussion relevant not only for architects and planners with their innovative understanding of cities in all their aspects, be it architecture or planning. Multiple contemporary projects start to grasp this understanding to different degrees, although not yet directly advocating the concept of a network city.

The main idea of a network city concentrates on network infrastructures – transport, telecommunications, capitals, energy, water, and streets – that form cities and urban areas, this way, contemporary urbanism emerges as a complex and dynamic sociotechnical process. Therefore, these infrastructures are viewed *"as ecosystems of competing networks"* (Graham S., 2000), which could be extended on landscape. Integration of information and communication technologies in urban areas is much more than a technological matter (Graham S., and Marvin S., 2001; Graham S., 2000). Considering the strategies adapted in Amsterdam and Barcelo-

---

41.
See: http://projekter.aau.dk/projekter/files/14420662/viby_an_interspace_pamphlet.pdf

na, it becomes evident that cities and landscapes aspiring to become smart should proceed with caution when adopting an approach that goes beyond technology, and consider other non-technical but yet crucial factors: human elements that are less discussed; collaboration between organizations of various sectors and citizens (Public-Private-People Partnership); long-term general goals and specific objectives; communication and promotion; availability of financial resources; ability to select the right combination of projects to develop in the short and medium terms.

As already mentioned above, a smart city is an integration of technological and social components, as well as an effective urban development model, that all together aim at more "intelligent'"and visibly visionary cities, "smart" not only from the perspective of using new and more innovative technologies but also because of collecting and mingling innovative actions and devices. In order to create these smart cities, as Maurizio Carta (2009) argues, we must adopt an urban life cycle planning and management model that is capable of constantly integrating ICT components with governance and local urban decisions. In this sense, we are living inside a "relational pathology" (Carta M., 2009) as a result of pseudo-integration between residential and production contexts, buildings and public spaces, knots and nets.

The above mentioned "relational pathology" resides in these places, and it is there that we should go first in order to find informed solutions that also take into account socio-economic changes in action. In this process, the era of networks, which we are witnessing in Europe nowadays, comes to our rescue, as it consists of nodes and streams of connections that keep cities linked through infrastructure, such as roads, railway lines, smart grids (material), and also energy, communication, Internet, and networks (immaterial). This will lead to creation of new interactions, new European contexts, and new paradigms.

First of all, in this new paradigmatic structure we find what in the present research work we call landscape 4.0, suggesting that landscape as a network is a place, in which "Smart Landscape" can already be deduced, seen and understood as a primary resource not only in terms of heritage but above all as a source of development.

IMG. 29
Book "Landscape Urbanism Reader" by Charles Waldheim

The resource of "Smart Landscape" is articulated as a network, which is linked, studied and correlated in different ways, and makes it possible to look at the future interests of a city and landscape, and to overcome a modernistic planning conception, according to which new physical structures create new forms of socialization. Rather, it is multiple urban dynamic processes that are poorly adapted to rigid and predetermined spatial forms. These dynamic processes, as a result of Landscape Urbanism in action, are able to extend to an intelligent network of related surfaces, or "Smart Landscape", in perspective of integration of advanced technologies into urban contexts in order to reduce the levels of pollution and consumption of soil and energy on our planet.

If we observe correlation between today's technological development and the definition of landscape 4.0 as a network of connections, then within this network we can identify an ecological matrix, which is also integrated into Landscape Urbanism and becomes a lens through which to analyze and to design future urban and landscaping improvements. This process of analyzing such fluid and non-linear organic development studies the way in which individual elements contribute to production of incremental or cumulative effects that continuously evolve in the environment over time. Therefore, seemingly incoherent or complex conditions may reveal a highly elaborate system structure: in this sense, cities and infrastructures remain as ecological as forests and rivers (Corner J. 2006). Landscape Urbanism guaranties development of space-time ecology that considers all the forces and agents working in the urban field as a continuous network of interrelations (Corner J. 2006). Kahn in his "Vehicle Circuit Diagrams" [42] (1952-1955) suggests contemporary techniques for the representation of such fluid processes. From "terra firma", therefore, we pass to "terra fluxus".

Landscape Urbanism projects and, in this case, articulation of a landscaping network, are characterized by a careful management of a project in order to pass to subsequent phases of territory transformation, which is clearly illustrated by the strategy designed for the Freshkills Park Landfill Competition (Staten Island, 2001)[43] by Field Operations, James Corner and Stan Allen. In this project, following an in-depth analysis of the existing human, natural and technological systems of the

42.

IMG. 30
Louis I. Kahn Traffic Study, project, Philadelphia, Pennsylvania, Plan of proposed traffic-movement pattern 1952

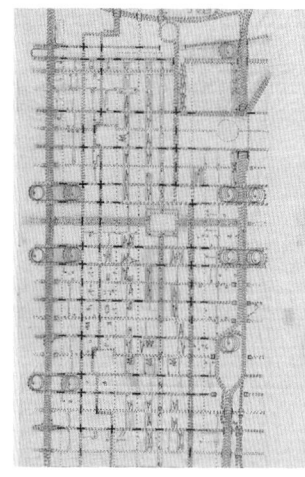

area, a series of overlapping maps and charts were organized into axonometric sections in an attempt to show the success and overlap of the activities in the context of reconstructing the site's ecological balance (Shane D.G., 2004).

43.
IMG. 31
Freshkills Park,
img. by PSFK

Landscape, in this case and from here on, can only be seen and considered as a system of infrastructure-technological, social-cultural, economic and information exchange networks. The nature of connections, influencing material conformation and specific operations on the territory determines the morphology and the way of occupying landscape of urban and suburban contexts (Lyster C., 2006).

New interchanging processes are in this case recognizable as a result of new practices that demobilize and relocate resources through and across a variety of sites and disciplines. From an object/site connection, the planning process evolves into an organization of higher complexity, which is created on multiple levels. From singularity to plurality –a territorial interpretation of new exchange processes is born out of this observation (Lyster C. 2006).

James Corner also resumes the concept of an exchange network system in "Landscape Urbanism, A Manual for the Machinic Landscape" by Mostafavi and Najle (2003). Every element of the system is connected to all the other elements through co-dependence and interactivity evolving soft systems that are sensitive to any change stimulus, which allows absorbing, transforming and exchanging information with what is around. Robustness and stability of this stimulus stems from its ability to handle dynamic differential motion processes (Corner J., 2003).

The way we see a territory is nurtured by an ecological interest in environmental sustainability, linked not solely to the balance

IMG. 32
Book "Landscape Urbanism a manual for the machinic landscape" by Mostafavi M., Najle C.

between produced and consumed energy, but also to the harmonious integration in time and scaling of rural and urban landscapes. The Landscape Urbanism approach, therefore, intends to develop a space-time ecology that includes every force and element operating on the field of action and considers them as a continuous and complex network of interrelations (Corner J., 2003).

Exchange networks that depend on simultaneous large-scale deployment over a wide field of play colonized territory that enables maximum mobilization as well as an amplification of exchange across other networks. (...) High-speed infrastructural networks such as distribution hubs, airports, rail yards, highway interchanges, and inter-modal facilities are natural collisive sites. Collisive territories tend not to occur accidentally but instead accumulate upon each other over time and are commonly coincident to pre-existing sympathetic conditions such as infrastructure, geographical features, natural resources, and program; that is, they reconfigure and magnetize existing topography rather than develop it from scratch. Exchange is thus multiplied and accumulated across multiple networks that occupy these collisive sites, suggesting a larger scale of transfer or inter-exchange at a specific moment in time. (Lyster C. (2006), p.227)[44]. As a consequence of such overlapping and arrangement, instead of creating a new topography, the existing one is reconfigured and magnetized.

Exchange processes multiply and accumulate across various networks that occupy these collisive sites, referring to a wider scale of transfer and inter-exchange at a specific moment in time. Organization of collisive territories happens due to multiple intersections of networks, which are both a source and a necessary condition for their operation, either as inter-exchange or collisive points. In fact, sometimes various unforeseen occupations of the territory, similar to the case of Sangatte's temporary refugee base, may happen as reactions to a particular confrontational context.

It shall be possible to observe collective sites that have a potential to magnetize urban contexts, territories and landscapes by working both locally and globally, with the logics of a network. This way, they function on different scales by tackling issues of redefining their contribution and ecological balance of landscape, in which once-naturally tamed

44.
Exchange networks that depend on simultaneous large-scale deployment over a wide field of play colonized territory that enables maximum mobilization as well as an amplification of exchange across other networks. (...) High-speed infrastructural networks such as distribution hubs, airports, rail yards, highway interchanges, and inter-modal facilities are natural collisive sites. Collisive territories tend not to occur accidentally but instead accumulate upon each other over time and are commonly coincident to pre-existing sympathetic conditions such as infrastructure, geographical features, natural resources, and program; that is, they reconfigure and magnetize existing topography rather than develop it from scratch. Exchange is thus multiplied and accumulated across multiple networks that occupy these collisive sites, suggesting a larger scale of transfer or inter-exchange at a specific moment in time. (Lyster C. (2006) p. 227)

nature now becomes wild again. These collective sites are changing the vision of planning into a stratified preparation involving countless social, public and private subjects present in the neighborhood and working in sensitive contexts of evolutionary processes. This action permits developing a territory through a differentiated action of interchangeable mobility as an attempt to harmonize interconnected infrastructure networks of different levels. The scale change is handled in a harmonious way by integrating the pre-existing fabrics, which implement the scale modification necessary for the urban operation of the flow space. At the same time, they promote the sense of belonging to bigger territories in the local context; this happens due to the temporary access to neighboring, far-reaching and distant points, i.e. neighborhood, city and region.

Later, the passage expresses an interesting exchange between the point of the network and the field, collisive sites and collisive territory or action field demonstrating virtuous involvement of various urban contexts. By exploiting its innate transdisciplinary ability to cross five dimensions of a smart city and to overcome them to become that city, it becomes both a paradigm of change, a code through which and with which one can design and relate to future projects, which will permit creating a Smart Landscape.

IMG. 33
By: Gustav Dusing
L.A. transportation network | Mapping 2016

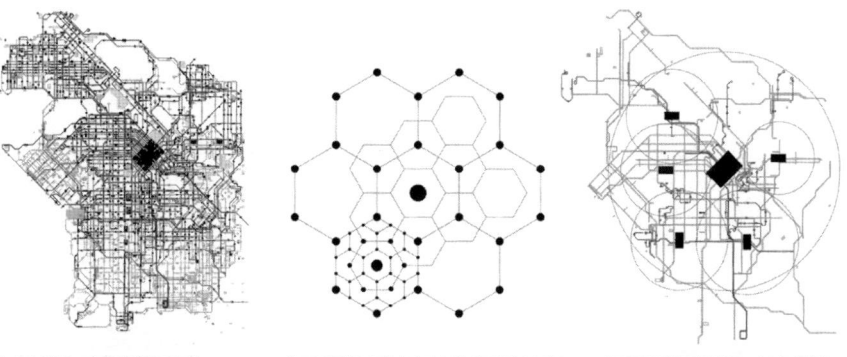

Los Angeles Public Transport_ the index of the decentralised city   The central place theory_ the hierachy of centers in within a bigger infrastuctural context   Implementing local hubs_ reducing driving ways / recreate a local identity

3. SMART LANDSCAPE

3.5 LANDSCAPE

The founding statutes of Western architecture have been questioned by the theories of Landscape and Ecological Urbanism for years (Mostafavi M., 2003; Waldheim C., 2006). The latter brought the discussion to a new level while dealing with apparently more fragile but certainly more flexible issues of formulating new approaches within landscape. Today, the disciplines of Landscape and Ecological Urbanism no longer deal with environmental issues only, now being of global interest, but also establish connections with the German aesthetic studies and focus on some specific conditions that respond to different problematic cores of fundamental interest.

As Elisa Cattaneo claims in her article "Landscape and Urbanism altered? Short circuits: the North American approach to urban transformation",

> *"... Landscape Urbanism and above all Ecological Urbanism seem to be characterized by some specific conditions ...:*
> *- Ecosophy as a holistic and all-encompassing system of research ...;*
> *- the interdisciplinary openness of research, to the point of making space sciences lose their focus, replacing their contents and tools for the benefit of a transdisciplinary approach;*
> *- the advancement of ecological logics to the detriment of typically urban logics ...;*
> *- overcoming of the dual/dialectic concept for the benefit of a dialog, in particular in the relationship between culture/nature and nature/city ...;*
> *- the leap in architectural scale and architecture in general as a topic of interest both for the transformation of the city and for the ecological question, to the advantage of the territorial scale and of the dynamic and procedural planning;*
> *- the advantage of horizontal surfaces over vertical ones ...; "*
> *(Cattaneo E. 2013)*[45]

45.
"... Landscape Urbanism e soprattutto dell'Ecological Urbanism sembra caratterizzata da alcune specifiche condizioni ...:
- l'Ecosophia come sistema olistico e onnicomprensivo della ricerca...; - l'apertura interdisciplinare della ricerca, al punto da far perdere centralità alle scienze dello spazio, sostituendone i contenuti e gli strumenti a vantaggio di un approccio transdiscipliare; - l'avanzamento di logiche ecologiche a discapito delle logiche tipicamente urbane...; - il superamento del concetto duale/dialettico a vantaggio di una posizione dialogica, in particolare nelle relazioni tra cultura/natura e natura/città ...; - il salto della scala architettonica e dell'architettura in generale come tematica di interesse sia per la trasformazione della città, sia per la questione ecologica, a vantaggio della scala territoriale e della pianificazione dinamica e processuale; - il privilegio delle superfici orizzontali su quelle verticali ...;". (Cattaneo E. 2013)

Out of these specific conditions and statements it should be clear that one of the greatest strengths of this discipline resides in its generality, not only on a conceptual but also on a scale level. Generality allows landscape to be 'combined' with multiple disciplines, lets it adapt to different contents and specifies its conditions. Due to this feature, many see and use landscape as a lens through which to design and/or rethink various places and contexts, thus representing the means by which to construct both cities and surrounding rural areas, going beyond the scale.

Currently, it can be stated that landscape has become articulated through a series of connections and networks. Therefore, "Smart Landscape" has its roots in this network. It is in this correlation between landscape and network that we find the answer to the environmental needs that occur in antropized contexts. To do this, we do not have to consider environmental and anthropized contexts as separate, which has been normal until now, but we have to work on landscape in an intelligent way, overlapping these two "layers" and trying to identify tools and techniques that could help in this process.

Right now, we are witnessing a 4.0 landscape, a technologically advanced landscape, which uses new technologies to its advantage and is taken as a slogan for industries or environmental policies. Nonetheless, landscape has not become 'smart' only for the mere integration of landscape and technological surfaces.

Landscape can be considered Smart when, by exploiting technological, political and social progress, we want to create a "responsive" landscape, where the actors, who build the places of living, production, agriculture etc. in a conscious way, understand that there is a need to work and build different landscapes. These landscapes, in a more or less symbiotic manner, should be able to use the most advanced technologies to become "sentient", i.e. able to recognize the limits of sustainability.

We do not wish to trivialize the strength of landscape or to follow a fashion, but we would either like to insist on the fact that landscape is the area to refer to due to its ecological sense. It becomes a system within which different contexts (for example, a city) can be placed. Within landscape we can propose an articulation of nodes that are able to become catalysts of regenerative processes both in the places in question and in the surrounding areas. Exploiting the large scale of landscape we can work with different surfaces, devices and policies, creating a sort of re-alignment, in which landscape replaces urban planning and architecture (Corner J. 2006).

IMG. 34
Book:
Detroit strategic framework plan December 2012.

Thanks to landscape, a mixture of a wide range of disciplines could be created. Thus, landscape could become an objective through which one could represent the diffused contemporary city and through which it could be constructed in relation to the contexts that are interposed between one city and another. The project for the city of Detroit, "Detroit Future City 2012", is an example of this propensity towards a "Smart Landscape". It has articulated the landscape of the emblematic city of Ford in a set of new infrastructures (blue, green and gray), which were able to act as an overlap of landscape layers, re-energizing the urban fabric with different initiatives in order to bring in new working places and residents, using an innovative approach to transform unused and/or abandoned areas. It promoted a diverse range of residential and sustainable housing units, and a management system for various areas of the city by balancing between different long-, medium- and short-term strategies. The project used landscaping tools as all-encompassing planning tools that could be divided into different networks on which to intervene individually to aim for a common goal. Particularly important in this project (and in the project of Freshkikks Park _ Landfild Competition (2001)) is its main goal characterized by successive phases of transformation of the territory.

In this research work, connection between smartness and landscape approach is present in various aspects: scaling, transdisciplinarity, connections between different systems, strength of its generality, etc., but above all in the change, which leads us to take action to anticipate and respond to it with smart tools, actions and processes.

This way we can argue that the strength of landscape associated with a smart system of connections such as energy smart grids (which act as a network) can lead to creation of a "Smart Landscape", which, remembering the affirmation of Michael Jacob, could be expressed in the following formula:

$$S.L. = S.G. + (S+N)^{46}$$

46.
Smart Landscape = Smart Grid + Landscape, re-creation of a formula introduced by Micael Jacob.

IMG. 35
Smart Landscape concept By Giulia Garbarini

3.5.1 LEARNING FROM THE STRATEGY OF SURFACE

Charles Waldheim maintains that today landscape architecture has become exactly what urban planning represented in the last millennium.

To start discussing and approaching the strategy of surface we must understand that we can no longer deal with and relate to landscape and urban disciplines as if we were living in the reflection of modern architecture.

Today, we must seize the opportunity to represent dispersed, immaterial urban and territorial conditions and therefore act with greater freedom. Within the disciplines of urbanism and landscape, introduction of Andrea Branzi's theory of *"weak widespread architecture"* leads to overcoming the existing dogmas and to turning our interest and our gaze towards minimizing the constructing limits and planning to turn them into quality producing energy that changes over time.

Today's urban and landscape condition of services and computerized, engineering and social networks leads to creation of a network of sensorial and smart "tunnel" systems, which cannot be represented and used with the classic architectural approach, but which must be entrust to a fluid planning and an elastic demarcation of what can happen in our urban or rural contexts.

A forerunner of all this is the "No-stop city" of the Achizoom, which illustrates an urban system without borders, without a clear distinction between what city and landscape represent, describing an urban context as a system of connections, forces and flows, which frees the modern society from the alienations that it had created (Branzi A. 2006).

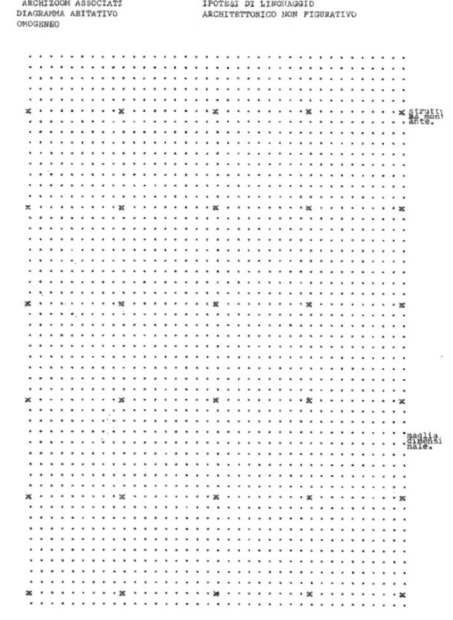

IMG. 36
Archizoom. No-Stop City, Residential Parkings, Diagram Homogenous Habitation, 1970

Andrea Branzi has worked for almost four decades to eradicate the intellectual poverty that for years had been accompanying Italian and international urban development. Thanks to him we were able to respond to this lack of intellectual conscience with "projects", such as "No-stop city".

In the interview in 2005, Branzi himself defined the "No-stop city" as a project/concept that had the ability to reduce and not to add or create, to arrive at a species of nirvana integrating different parts of a cosmos understood as urban and landscape environment[47]. Another symbolic project is Acronica, in which we see urban planning as an economic and aesthetic environmental form created of weaknesses of a contemporary city, which makes us rethink driving force as a "strategy of surface". The "strategy of surface" is born and developed to be applied

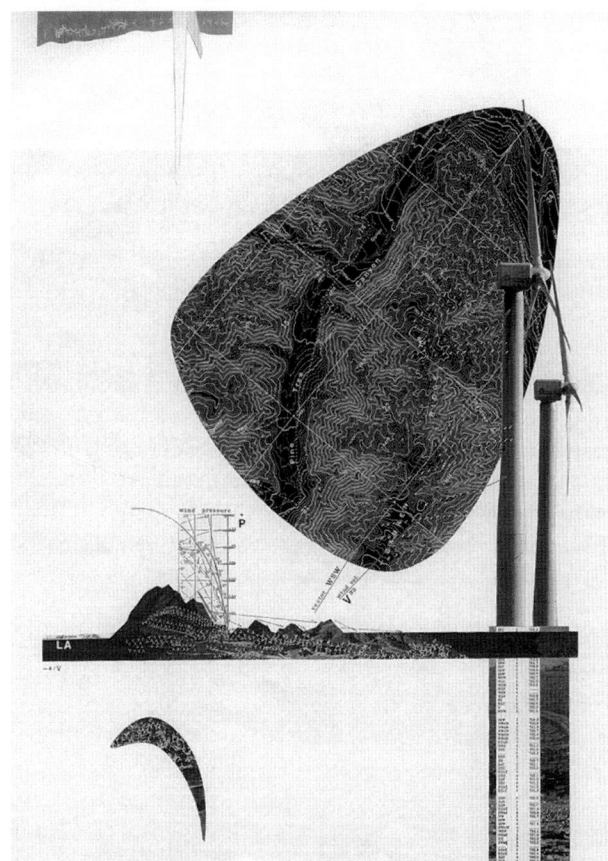

IMG. 37
Drawings by James Corner and Alex McLean. In: "Taking Measures Across the American Landscape"

46.
Smart Landscape = Smart Grid + Landscape, re-creation of a formula introduced by Micael Jacob.

47.
"Andrea Branzi: la ville continue [Interview with Andrea Branzi]", article by Antonio Scarponi in Moniteur Architecture AMC no.150 March 2005 / p.88-94 (text in French)

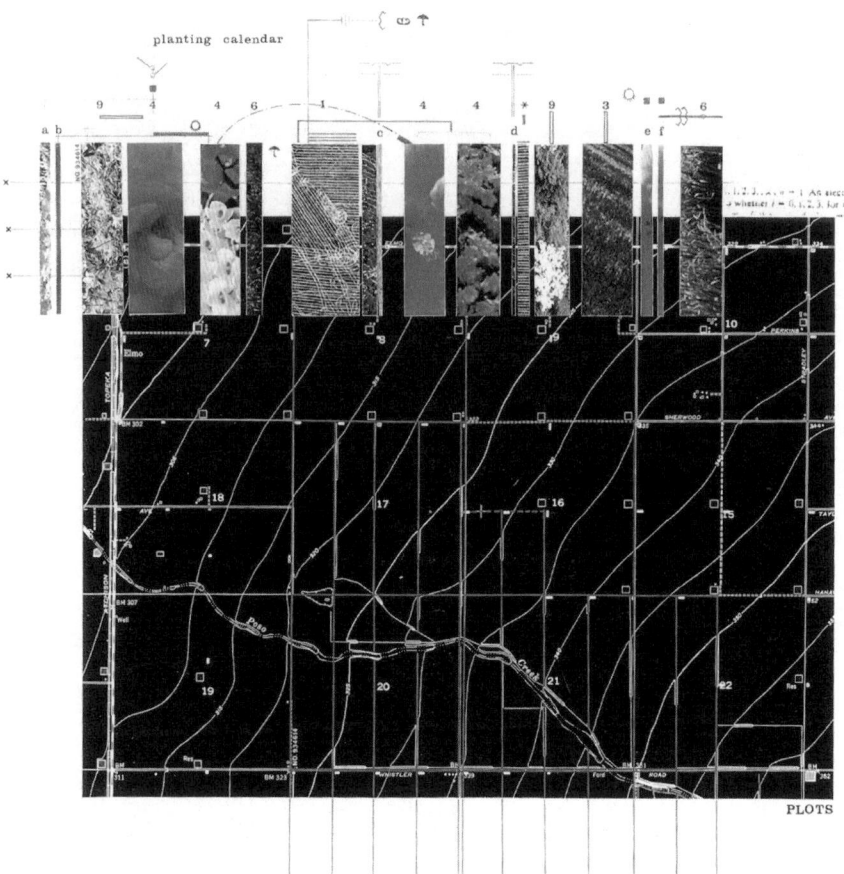

IMG. 38
Drawings by James Corner and Alex McLean.
In: "Taking Measures Across the American Landscape"

to landscape context, which not only works or is identified within a territory to be preserved and protected but which intends landscape as a "surface" without borders.
In fact, landscape designer and teacher Jeams Coener explains this theory: *"Land division, allocation, demarcation and the construction of surfaces constitute the first act in staking out ground; the second is to establish services and pathways across the surface to support future programs; and the third is ensuring permeability to allow for future permutation, affiliation and adaptation" (Corner J. 2003, p.60)* [48].
The most interesting aspect of this theory is its approach to the context. In fact, it is characterized by two phases that work simultaneously on the same site, incorporating a perceptive

48.
See James Corner's Article "Landscape Urbanism". In Mostafavi M. & Najle C. (eds.) "Landscape Urbanism: A Manual for the Machinic landscape." AA Publications p. 58-60

approach aimed at understanding the characteristics of the site and at the same time a capacity to adapt to its future needs. This theory elaborated by James Corner can be seen, especially in the present research work, as a methodological base on which a new concept of landscape (urban) planning will be erected. It was conceived to satisfy a growing demand for new planning practices and therefore for landscape and urban processes that take the place of the old ones, now obsolete and no longer able to respond to and manage the growth dictated by globalization.

The challenge proposed by Corner (Terra Fluxus) in the book "Landscape Urbanism Reader" by Charles Waldheim resides not only in bringing landscape into the urban system but also in expanding the city to the surrounding landscape. Before, they were considered separate but now, more than ever, they should be seen as one. This landscape theorization might appear to be incoherent or too complex, but, as mentioned before, landscape can be seen and shown as a comprehensive tool, a set of geometric and spatial instruments that intend using urban systems and related infrastructures as an "ecological" tool similar to rivers and forests (Waldheim C. 2012)[49]. Hahn, with his proposal for the "Market Street Eat", shows us the need for efficient diagrams and tools to represent the flows that characterize the contexts. In this relation, Corner identifies the need to use an approach linked to the horizontal alignment of landscape, calling it "horizontality".

In this "horizontality" reside the three layers of horizontal surface theory, which makes part of the "strategy of surface". The first two layers represent "demarcation" and "infrastructure", still linked to conventional planning, while the third one ("adaptation") is seen as an additional dimension of the conventional

---

[49].
In conceptualizing a more organic, fluid urbanism, ecology itself becomes an extremely useful lens through which to analyze and project alternative urban futures. The lessons of ecology have aimed to show how all life in the planet is deeply bound into dynamic relationship. ...in fact be shown to be highly structured entities that comprise a particular set of geometrical and spatial order. In this sense, cities and infrastructures are just as "ecological" as forests and rivers". See James Corner's Article "Terra Fluxus". In Waldheim C. (2012). "The Landscape Urbanism Reader". p. 23-33

IMG. 39
Parc de la Villette OMA proposal diagrams, 1982. Img. by OMA Studio

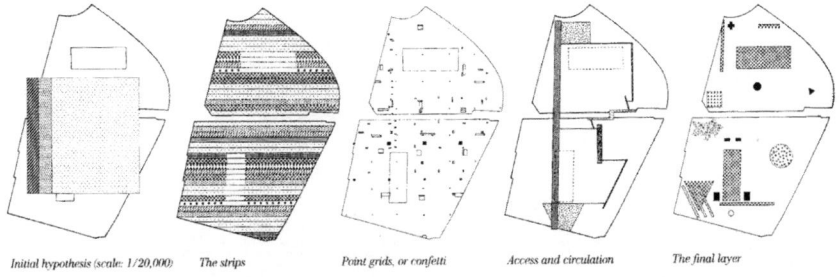

Initial hypothesis (scale: 1/20,000)   *The strips*   *Point grids, or confetti*   *Access and circulation*   *The final layer*

3. SMART LANDSCAPE

approach. Adaptation incorporates both ecological and dynamic processes of development that require and allow us to create a more flexible planning.

This layer – "adaptation" – is fundamental for the present research work because it supports our concept of scaling landscapes, which, as in the case of this third conceptual layer introduced by Corner, characterizes a new inherent approach in Landscape Urbanism theories. This approach is articulated differently in its procedures and timing depending on the process and objectives that are to be achieved in a given context. The strength of adaptation is therefore in its ability to change over time according to the needs, related to the other two layers. This fact allows us to work on a project in a planned way without compromising on the general concept as a result of connections that "demarcation", "infrastructure" and "adaptation" incorporate and activate.

Corner hypothesizes that in his three surface strategies there is a new way of incorporating and working with the dynamics of a contemporary city. For example, if we understand infrastructure as a broader definition of roads and paths, they imply a perception of infrastructures that operate on a site as both tangible and intangible (such as communication between different nodes of an intelligent network). In this perspective, the second of the surface strategies and its processes are inserted in thenew matrixes of interconnections with the ability to incorporate new links into the site returning to the concept of infrastructure. As stated by Rune Bach, in his doctoral research on the surface strategies of Corner, the third layer – "adaptability" – must be incorporated into "demarcation" and "infrastructure" because it introduces permeability and adaptability in relation to these two (Bach R.C. 2008).

The strategies of surface prove to be very demanding for conventional planning, that is why we no longer need to base our work or design on rigid masterplans typical for the modernist period. On the contrary, we must direct our gaze towards more dynamic plans that are recognized and articulated by progressive layers of the existing site, working with the potential stimuli of each layer and creating new ones based on the contest date. The first example of working with layers was the project for the Park of the Villette in Paris by OMA – Rem Koolhaas.

It was aimed at achieving a greater flexibility in the totality of the park and in its planning over time. The areas and surfaces hypothesized by Koolhaas adapt to the temporal and physical changes of the city and the users needs due to the absence of a unique infrastructural program, which guarantees considerable flexibility.

In the present work, we tried to recreate this flexibility by mingling technological instruments, such as smart grids articulated by connections of nodes, and landscape, which has both a noticeable transdisciplinary potential and the ability to work with surfaces as layers of a system.

3.5.2 DRIVERS TOWARDS A SMART AND ADAPTIVE LANDSCAPE

The European Union, like other major industrialized economies of our planet, has committed to supporting a massive integration of numerous actions and devices with a view of achieving the international goals of reducing carbon gas emissions, disasters correlated to the climate change etc. Critical awareness of these goals remains fundamental, and the need to take adaptive actions in the energy, society and environmental fields is recognized. Climate changes underway, along with their inevitable impacts and consequences not only in the environmental field, suggest that the appearance and the spatial organization of urban and rural landscapes will undergo influences and far-reaching changes. Since one of the fundamental objectives of our era is creating resilient production of renewable energy, management of resources on our planet and mixing of social, environmental, and energy streams, the task is to reintegrate them into the landscapes in which they are located or created, and thus make a qualitative and quantitative contribution.

In order to arrive at the definition of adaptation and smart strategies that are capable of meeting the European Union goals, we should be able to become aware of the limits of certain disciplines, and also to understand how their interaction can benefit from improving bad conditions.

This chapter and in particular this paragraph would like to show why it makes sense to chose the discipline of landscape to deal with such complicated topics that have as a goal safeguarding our planet and making our urban and rural contexts resilient and adaptable.

One of the motivations is found in the book *"Climate-Smart Landscapes: Multifunctionality in Practice"* edited by Peter A. Minang, Meine van Noordwijk, Olivia E. Freeman, Cheikh Mbow, Jan de Leeuw, Delia Catacutan.

In the first chapter of the book, landscape is described as a

> *"mosaic of different land uses with multiple components and functioning interactions between and across ecological, social and social-ecological process (functional interaction), made up of multiple actors and stakeholders with varying interest (negotiated scales), and made up of nested components occurring on different scales (multiple scales)" (Minang P.A. et. al., 2015, pp.3-8).*

This definition is shared in the present research, because it expresses the transdisciplinary nature of landscape and creates a matrix of recognized units such as rural areas, communities, cities, productive surfaces, energy and ecosystems. (Mentioned in Minang P.A. et. al. 2015: Parrott & Meter, 2012).

Plans, projects and devices used in the case study of this research are chosen as part of the "mantle of landscape". The idea would be to understand how this work could lead to development of landscape as a "project" by interpreting the concept of adaptation as a process of anticipating negative effects of climate change and as a "design" of action. The purpose of this action is to prevent and reduce damages caused by the climatic changes that represent our era. The present research studies and adapts the philosophy of Climate Adaptation Plans, which could become real project plans. These plans incorporate various tools and strategies to make a city and adjacent territories adaptive to future changes (European Commission, 2007). Their realization allows consolidation of the concept of "pro-activity" targeted on simultaneous processing of plans that work on the energy issue and provide for productivity and smart storage.

The devices and tools taken into account in this work were selected in order to launch a complicated dialogue on the issues Architecture and Urban Planning are challenged to address nowadays. These issues are far too complex to be investigated and dealt with only within the borders of these two disciplines. This brings up a question: how can we deal with the complex transitions that will be faced by territories and cities of the future?

The current tendency within the European geopolitical framework is to direct theory and design towards a "post-fossil" (Sijmons D., 2014) context. This goal is distant, but it can be attained gradually through a smart and technologically advanced process. It arises out of the need to detect and use appropriate policy-planning strategies in landscape and urbanism. These strategies spread from energy markets to legislation, from the concept of adaptation to technological development. We should therefore be allowed to operate in a transdisciplinary approach, which can bind and develop technology and urban design at the same time (Brundtland, G.H. 1987).

With the help of this transdisciplinary approach appropriate answers to various questions can be found. This approach, now called Landscape Urbanism, emerged as a practical and theoretical concept 16 years ago, and since then numerous members of the scientific community have been exploring it as an extremely convincing idea. Landscape Urbanism is transdisciplinary by definition, because it brings the legacy of landscape design to the contemporary urban dynamics; it units environmental engineering knowledge and techniques, urban strategy and ecology of landscape, deploying the science of complexity and emergency to the tools of digital design and political ecology. Combination of all these techniques and knowledge forms a new tool for project interventions in urban planning seen as a social, material and environmentally friendly factor continuously modulated by spatial and temporal forces and connected into a network.

The present chapter, as already mentioned above, is intended to attract attention to this transdisciplinary framework, which resides in the very concept of landscape and is directly connected to urban design, renewable energy, and their local (urban and district) energy network systems. This connection facilitates inclusion of landscape into the present phenomenon of hyper technology. It becomes possible due to a new perspective, according to which the most technologically advanced disciplines are connected with those (including landscape) that have learned to be resilient to changes and have become the central point in the contemporary debate. This process is happening within a system of interconnected networks, which become invisible infrastructures. The intent is to make use of the models and tools that engineering disciplines develop to meet and respond to climate change and to transform them into instruments that pro-activate high-quality urban, rural and social landscapes.

A recent research connected to these reflections permitted the author to develop a line of reasoning using (micro and macro) smart grid[50] as an architectural design model (AA. VV. 2011). Renewable energy production plants (for example, wind turbine generators) were used as tools and considered elements of urban and landscape processes due to their various possible applications. The idea was to combine a distribution model (smart grid) and renewable energy production tools. It derived from a general reflection on the European Strategic Energy Technology Plan (SET-plan)[51] and European Landscape Convention (Council of Europe, 2000). The first promotes energy policy that indicates technological change as part of the reconfiguration of the living space of Europeans, and the second interprets transformation as one of the ways to improve the characteristics of landscape (Morata F., Sandoval S. 2012).

This reasoning creates a decent margin for a possible integration of landscape with energy policies. This integration goes beyond the instrumental dimension of landscape and leads to creation of a more beneficial system of energy/landscape policies. Committee of Ministers'[52] recommendations (CM / Rec (2008) to the member states on the guidelines for implementation of the European Landscape Convention introduce the idea of integration of landscape with sectorial policies and thus with energy business (Morata F., Sandoval S. 2012). Considering the latest meeting on the EU's climate change and energy policy objectives for 2020 and beyond, the current situation will require a major transformation of the electrical infrastructure (Camarsa G. et. al., 2015).

50.
The Smart Grids Task Force advises the European Commission on the development and deployment of smart grids.

51.
The SET-Plan, adopted by the European Union in 2008, is the first step to establish an energy technology policy for Europe. It is the principal decision-making support tool for European energy policy.

52.
The CM / Recommendation Rec (2008) 3 of the Committee of Ministers to member states concerning the guidelines for the implementation of the European Landscape Convention (adopted by the Committee of Ministers on 6 February 2008).

It is extremely important to strengthen and upgrade the existing networks in order to increase the amount of renewable energy production. This way, grid security, development of the internal energy market, significant energy saving, and efficiency of the system could be enhanced. To achieve these goals it is not only necessary to build new lines and substations, but it is essential to make the overall electricity system smarter through integration of Information and Communication Technologies (ICT)[53]. Smart Grids can be described as an upgraded electricity network enabling two-way information and power exchange between suppliers and consumers due to incorporation of intelligent communication, monitoring and management systems. In the last few years, initiatives on Smart Grids (with different aims and results) have been growing in number and size throughout Europe.

By analyzing these strategies and technological models designed to address the questions of our future from the point of view of landscape change, will it be possible to integrate urban or rural landscape into energy policies with the use of urban, landscape and social devices?

The answer is "yes". If we can imagine this happen, we will not only be able to offer technical benefits but also help cities become more sustainable and attractive. Smart grids could deliver power infrastructure, which is convenient, reliable and efficient. They could facilitate and support sustainable growth processes allowing urban planners, designers and landscapers to make the most efficient use of the potential of lanscape through exploiting a model of interconnected networks that will become a polygon for creation of places - catalysts of values.

The future transformations projected from a sustainable point of view will change the smart grid model. For example, a wind power generator can become an integrated tool that is compatible with new uses and community service operations. Thus, this particular approach to transitioning becomes the best practice of participation, management and performance control over time. The first goal is to engage further compatible and enabling technologies that are capable of integrating innovations (ICT) in terms of "smart communities" to obtain the best quality of life of the site residents and visitors. The second is to regenerate spatial configurations on urban and

---

[53.]
ICT (information and communications technology - or technologies) is an umbrella term that includes any communication device or application. According to the European Commission, the importance of ICTs lies less in the technology itself than in its ability to create greater access to information and communication in underserved populations. Many countries around the world have established organizations for the promotion of ICTs, because it is feared that unless less technologically advanced areas have a chance to catch up, the increasing technological advances in developed nations will only serve to exacerbate the already-existing economic gap between technological "have" and "have not" areas.

rural landscapes and to improve environmental conditions and energy consume with the help of a "smart" and "green" model. Wind energy has become a particularly controversial issue in many communities, which have been weighting the pros and cons of its use. In some places, this leads to formation of local, yet highly polarized opposition and support groups, which tend to exclude neutral, indecisive, or cautiously opinionated individuals. At the same time, wind turbines are present in more than 80 countries today. They emit no greenhouse gases, and therefore reduce the climate change consequences. As showed in the IPCC Special Report on Renewable Energies, wind turbines do not emit air pollutants, which cause acid rains etc., and in particular they can be installed in both rural (as plants) and urban (in parks, homes or public spaces) contexts. Various trims of interconnection between different transdisciplinary approaches will lead to innovation of "smart grids". This will give life to a different vision of the wind potential and creation of nodes (catalysts)[54] (Oswalt P. et. al., 2013) with the help of connections in the network. These connections will creat a different kind of urban and landscape design, which can be activated according to intervention and study scenarios, specific for different places of interest.

54.
The difference between the catalyst and these redevelopment strategies is that catalytic redevelopment is a holistic approach, not a clean-slate approach, to revitalizing the urban fabric.

IMG. 40
"Urban Catalyst: the
prover of temporary use."
By P. Oswalt. K.
Overmeyer. P. Misselwitz.

CHAPTER FOUR

REFERENCE

The present chapter intends to explore the idea of Smart Landscape through a selection of projects aimed at implementing today's mapping of information while trying to facilitate the process of cultural contributions both in landscape and technology. The latter includes systems used to apply smart grid technology and landscape projects, which through various initiatives have already reactivated areas of their location, and, therefore, could be applied through linking landscape and technology of smart grids.

The intention would be to put urban and landscaping projects under lens in order to understand their present and missing potentials as well as breakdowns, and to include them into the concept of "enabling technology" described by Professor Consuelo Nava[55].

The choice of selected projects should not be seen as a catalog of "technological prostheses" of a city/territory/landscape but as a set of experiments that in a new and integrated way can deal with a series of issues and needs. They work on technological dimension of participation, where citizens themselves provide useful information through taking part in the application of a project.

This process permits adopting certain devices that serve to improve the above-mentioned contexts and to create Smart Landscapes.

Reasoning in these terms and referring to what was said previously, advanced technologies are used to respond to environmental and social issues. They experiment with new forms of management and connection but above all work in terms of smart community acting through merging of landscape, tangible and intangible infrastructures, and humans towards creation of a univocal meta-organism.

This meta-organism, as mentioned above, can be considered landscape. It includes all the five principles of a Smart City and in behalf of the "European Landscape Convention" acquires a wider value, emerging as a strategic element on which to focus during the development of some "smart" categories postulated by the European Commission for the goals necessary to ensure protection of the inhabited areas. Among these goals are Smart Energy, Smart Community, Smart Mobility, and Smart Grid.

55.
Cft. Nava C. (2016). "The laboratory city. Sustainabe recycling and key enabling technologies" , edited by Aracne in Re-cycle Italy, Roma.

Therefore, this chapter intends to enclose and express a thought-provoking dialogue between the actors of the network (represented by innovative and resilient projects), collective territory, and technological advancement due to the use of smart grid technology, which becomes a polygon on which to implant a vision of Smart Landscape.

Starting with the above-mentioned assumption, this paragraph primarily describes the Smart Grid technology and deepens the research of projects/systems with the help of synthetic design schemes present and operating in Europe.

Subsequently, this chapter intends to remain in line with the conceptual approach of the present work, looking at contexts in which the technology of smart grids is present, where projects, plans and/or definable actions have already been activated, providing reading schemes that describe characteristics and potentials of these projects.

In the end, four projects were chosen. Each of them in the best possible manner expresses one of the four smart categories needed to obtain a Smart Landscape (see above). At the same time, even if in a reduced way, each project contains all of the categories, and, therefore, is intended as a successful reference to which the concept of Smart Landscape could be applied, interconnecting existing features and devices.

The experimentality of the Smart Landscape concept is to be outlined through combination of theoretical elaboration of the Smart Landscape model (see chapter 3), the study of reference projects both from technological and urban/landscape points of view, and through elaboration of the main prototype (case study of the Venice Lido) illustrated in a dedicated book[56].

---

[56]
Cft. Garbarini G. (2018). "L.I.D.O. - Learning Island Design Opportunities", edited by List Lab, Treno-Barcellona.

4.1 TOOLS AND DEVICES

Within this research work, like in numerous contributions both in present and other disciplinary fields (on the Internet, television etc.), various terms such as "tools" and/or "devices" are used to indicate technologies, technological innovations and objects indispensable in our daily life. They are generally aimed at satisfying a need or at simplifying urban and domestic life of users. These tools and devices aim to improve society, cities and landscape progressively.

> **Tool:** *a piece of equipment, usually one that you hold in your hand, that is designed to do a particular type of work (kitchen/gardening/dental tools); something that you use in order to perform a job or to achieve an aim; someone who is used by another person or group, especially to do a difficult or dishonest job.*
>
> **Device:** *a machine or piece of equipment that does a particular thing; a formal way of making something happen or of making someone do something.*
>
> *(Macmillan Dictionary)*[57]

The definitions quoted above only partially allow us to understand differences and similarities between the two terms. Obviously, they vary in relation to different disciplines, of which only some are of our interest and serve to understand what meaning is assigned to them within the present research work.
To understand the theoretical and design choices (that will be described later) it is important to determine what the present research intends for tools and above all for devices. This will also lead to understanding of how the latter could be used in urban and landscape contexts.
Although these two terms are quite inter-exchangeable or assimilated, it is natural to suppose that there is still a substantial difference between them. A tool is a utensil, a component

[57] https://www.macmillandictionary.com/dictionary/british/tool_1, consulted September, 20. 2017

as such that can be used to accomplish a specific action or makes part of a bigger mechanism. A tool can be used to create a device and can become an integral part of it, or it can be multifunctional.

Within the disciplines of architecture, urbanism and landscape, diverse tools are used to create a project, and the more technology and anthropology develop, the more various tools evolve, allowing creating and elaborating objects unthinkable in past.

In this regard, the book "Innovations in Landscape Architecture" (Anderson J.R., Ortega D.H., 2016) is of outmost interest. It addresses the issue of increasingly emerging and massive use of tools such as digital software and products of technological manufacturing inside landscape disciplines. The abovementioned book allows us to understand very clearly in what way past and ongoing evolution of these tools has played and will increasingly play a central role in the practice of landscape design, architecture and urban planning. In particular, this work highlights the use and application of digital tools in these disciplines, and the way their rapid and striking evolution has created avant-garde instruments for professionals, academics and projecting groups. Among those who were called to participate in creating the book "Innovations in Landscape Architecture" the contribution of Jose Alfredo Ramirez and Clara Olòriz Sanjuàn is of considerable interest. They reflected on the role of a designer who faces this superabundance of tools and, therefore, possibility of using them for the most varied interventions, offering both to users and "spectators" some new perspectives of application and vision of the contexts in which they are living today.

As J. A. Ramirez and C. O. Sanjuàn mentioned (2016, p. 9-27), *"the new technologies constantly change and re-shape the way we think, design and produce our environments and territories".* Numerous innovations in technologies which are now widely available to architecture and landscape designers were created as a result of a human impulse to control the surroundings in which he is immersed.

Numerous methodologies and study subjects based on innovative tools imply abstract systems of organization that provide frameworks to develop and apply concrete interventions and management schemes on given territories. Following the reasoning of the two authors regarding the role of digital tools in relation to a designer's figure we are forced to reflect on the role of a designer and his capacity to engage with territories and the dynamics that shape them using new tools at his best. As J. A. Ramirez and di C. O. Sanjuàn said, *"... digital tools can exacerbate designer's detachment from contemporary conditions (as a mere observer) whilst diminishing his direct participation and implication in the reality".*

In order to avoid what was previously prognosed by J. A. Ramirez et. al., we could consider that tools, as already mentioned above, are able to generate devices, which from a philosophical point of view can be considered as a *"skein, a multi-linear set, composed of lines of different nature... subject to variations"* (Deliuze G., 2007). As devices are a tangle with undefined contours, which is articulated by a concatenation of variables, it is evident that they become stronger during crisis, integrating and adapting to the context in which they are inserted and modifying their own nature with respect to the needs that they are supposed to satisfy. A device dissolves in response to a specific need, as it is *"the skein of the lines of the device itself... tracing a map, measuring unknown lands... in short... conducting field research"* (Deliuze G., 2007, p. 11-12). These philosophical statements highlight the fact that the concept of devices is different from that of tools.

In fact, the former, being *"machine or piece of equipment that does a particular thing"* relates differently to the context and its network of infrastructures even without help of humans. Tools are often referred to as "smart devices", electronic gadgets that are able to connect, share and interact with their user and other smart devices.

Devices that are often defined as ICT (Information and Communication Technologies) can be considered as part of these "smart devices". The latter have spread massively in cities and anthropized and rural landscapes in an unprecedented way. These devices, depending on their influence on social contexts, are classified (A. A. Abdel-Aziz et al., 2016) into four main categories: Wi-Fi networks, digital interactive multimedia applications, public interactive displays, and smartphone applications in public spaces. They become more and more connected to public spaces and in relation to various devices are divided into five main domains: culture/

arts, education, planning and design, games and entertainment, information and communication.

Based on this classification, various examples of ICT devices are present in our landscapes in relation to public spaces and buildings. In fact, by simply looking around it can be immediately noticed that digital technologies are present in almost any environment or context of daily life.

Since technology, applications and services are constantly evolving at a quick pace, we should keep in mind the current trend of promoting virtual interaction with direct contact, trying to integrate this technology into architectural, urban and landscape design. As stated by Areti Markopoulou (AA. VV., 2015), who believes that one of the paradigms of change in above mentioned disciplines (as they by nature cannot be disconnected from general changes) resides in integrating and trying to unite technological evolution with urban and landscape architectural design.

Landscapes of our times and of the past can be used to ameliorate all the assembly spaces within a community: streets, sidewalks, parks, buildings, and other public spaces. This idea is not new: it was born in the 1960s, when such visionaries as Jane Jacobs and William H. Whyte offered innovative ideas in order to design cities for people, not cars (Abdel-Aziz A. et. al., 2016). Today the goal should be to support and accompany creation of places of interaction between people and to push communities to become more socially, physically and economically sustainable (Anderson E., 2013).

In short, the devices and tools demonstrated in this work and in its prototype can be identified as Enabling Technologies, which have a considerable importance within the European policies. In this context, they are defined as Key Enabling Technologies (KETs).

[58.] European Commission / Growth / Industry / Industrial policy / Key Enabling Technologies.

See https://ec.europa.eu/growth/industry/policy/key-enablingtechnologies_en

"KETs are a group of six technologies: micro- and nano-electronics, nanotechnology, industrial biotechnology, advanced materials, photonics, and advanced manufacturing technologies. They are applied in multiple industries and help tackle societal challenges. Countries and regions that fully exploit KETs will be at the forefront of creating advanced and sustainable economies".[58]

Nonetheless, the definition provided by the European Union turns out to be unproductive considering that the aim of this research is to introduce a different way of operating on landscape. For example, *"... "enabling technologies" are articulated in their connection with "time", which is a condition of process and project that has always distinguished the effectiveness of sustainable actions..."* (Nava C., 2017).

As Professor C. Nava states and illustrates in her work, the devices that are intended to be used both theoretically and experimentally in this research, are considered as "enabling" technologies aimed at triggering more extensive and complex innovation processes. Due to increasingly high performance and commingling of several tools these devices aim at becoming part of a wider system, which brings additional value to the existing system while innovating the entire process.

The devices that will be mentioned in this work and later used in the case study (prototype) were selected due to their main feature: to include a certain project and its context into a process of evolution not only from technological but also from social and environmental points of view trying to enter into a network of innovative actions.

4.2 REFERENCE NETWORK

In the previous chapters, the term "network" was frequently used in association (in a new paradigm conformation) with landscape, which in the present research work is identified as "landscape 4.0". That is, landscape as a network, which is seen and understood as a primary resource not only in terms of being preserved, but above all as a source of development.

Thus, there is an attempt to identify and understand devices or tools that help adapt an approach that is "smart", innovative, avant-garde, and find sustainable solutions, which would take into consideration the ongoing socio-economic changes. The projects to which it was decided to refer were supposed to be contemporary to today's networks that we are now witnessing, living and planning in a Europe made of nodes and flows, which maintain connections between cities through material infrastructures (roads, railway lines and smart grids) and intangible ones (energy networks, see : communication networks etc.).

Therefore, the present research work and the related case study, which will be illustrated below, take their origins both in the imminent context and in a set of rules proposed by the European Union for smart grids, and intend to introduce a sustainable city model that can optimize distribution and consumption of energy while reducing its environmental impact.

Subsequently, we will illustrate the reference network (smart grids), its components and projects, which we analyzed and identified within the European context during the PhD research year.

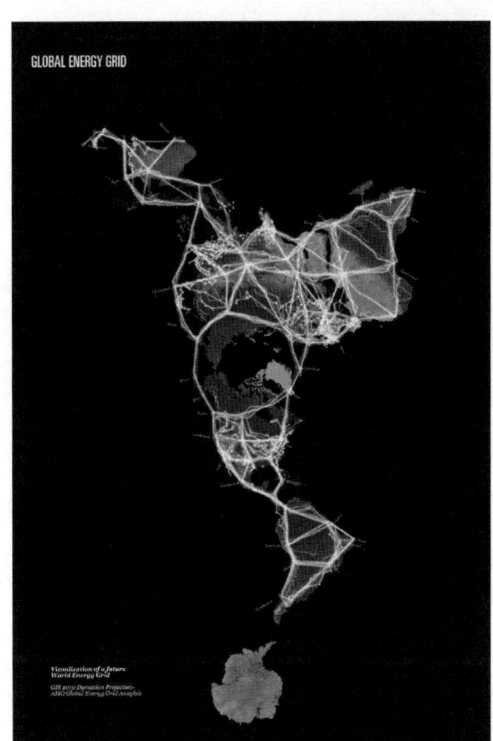

IMG. 41
Global Energy Grid illustration for Roadmap 2050.
By OMA/AMO studio

4.2.1 SMART GRID

"Smart Grids[59] could be described as an upgraded electricity network to which two-way digital communication between supplier and consumer, intelligent metering and monitoring systems have been added. Intelligent metering is usually an inherent part of Smart Grids."
(European Commission, 2011)

Smart grids could be considered as a new level of the current electricity distribution networks in relation to the growing demand for energy, expansion of renewable energy sources and evolution in information technology. They become an efficient solution from energy and economic points of view.

The rapid expansion in the use of renewable energy sources is most apparent in the wind and solar energy fields. For example, it is estimated that in the United States the wind energy sector will grow from 31TWh in 2008 (or 1.3% of the needs) to 1160TWh in 2030, which would be equivalent to reaching the level of 20% of total production (equal to about 5800TWh) (Wiser R., Berkeley L. et al., 2008).

Nonetheless, it is well known that renewable energy sources are unpredictable due to the fact that their production capacity is lower than that of conventional generators, in particular when speaking about wind energy. Strong variability and inconstancy in wind production is caused by the specificity of production equipment and its geographical distribution. The latter is non-linear, as large-scale wind sources are normally placed at a major distance from loads, leading to voltaic and thermal transmissive limitations, and above all to instability issues. As for solar energy, it turns to be the most abundant source of renewable energy, as the total solar energy that reaches the surface of the Earth in one year is about 1000 times superior to the world consumption of fossil fuels in one year (Kimbis T., 2008). However, like in the case of wind energy, solar energy presents the same transmission limitation due to distance between sources and loads. Ultimately, renewable energy sources are an important energy resource, but their use requires elasticity and capacity to adapt to factors of variability, which current electricity grids are not able to satisfy.

59.
The European Smart Grid Task Force defines Smart Grids as electricity networks that can efficiently integrate the behaviour and actions of all users connected to it – generators, consumers and those that do both – in order to ensure an economically efficient, sustainable power system with low losses and high quality and security of supply and safety. http://ec.europa.eu/energy/gas_electricity/smartgrids/doc/expert_group1.pdf. Consulted May, 20, 2017.

IMG. 42
The Global Smart Grid Federation 2012 Report

IMG. 43
Smart Grid projects in Europe: lessons learned and current developments.
By: Vincenzo Giordano, Flavia Gangale, Gianluca Fulli (JRC-IE) Manuel Sánchez Jiménez (DG ENER)

Growing demand for energy has stimulated creation of plans that tend to expand and upgrade existing electricity grids in industrialized and developing countries, raising the issue of growing economic and environmental costs that make it difficult to use old paradigms for such enlargements and upgrades. At the same time, continuous progress in ICT (Information and Communications Technology) has created a convergence of scientific and industrial interests in the use of these technologies that implement a process of structural transformation of each of the energy cycle phases: from production and accumulation to transportation, distribution, sales and intelligent consumption of energy. New functionalities are then activated or improved within the management of electric networks while transporting information flows that are necessary for specific tasks with the help of the current infrastructure and transforming the nodes of the network into active nodes. This combination of ICT and energy is commonly identified as "Smart Grid".

One of the major advantages introduced by Smart Grids is their ability to efficiently integrate renewable energy sources that are generally intermittent due to their dependence on non-constant phenomena, but that constitute an important energy supply if properly exploited. At the same time, if integrated in a correct way they can compensate for these shortcomings of constancy. The basic functionality of the smart grid system focuses on integration of distributed energy resources (DER) into the current system.

The key feature of smart grids is the ability to manage (through protocols and information flows) generators and active loads available within the network, coordinating them to perform certain functions in real time.

In this way, devices of an electric network become active part of the control cycle extended to large power stations as well as to devices of individual users. Moreover, passing on to a distributed generation system of this type, transmission losses are greatly reduced since electricity is produced in large part where it is consumed due to the use of renewable energy sources as generators connected directly to the user. A fundamental consequence of the use of information that flows through the current infrastructure of a network is a possibility of managing peaks of maximum usage through scheduling of loads in order to avoid their simultaneous activation. One of the critical aspects in energy distribution today is the existence of peaks of usage. During these peaks, auxiliary standby generators are used to ensure a constant energy flow without interruptions. They are activated in order to avoid voltage drop due to numerous network loads that occur in the same moment. It is clear that generators of this type constitute a burdensome economic resource, and their elimination would significantly increase overall efficiency of the system. These peak loads can be reduced by implementing some consumption regulations combined with the use of smart meters (digital meters that communicate with the rest of the network) and automated load management systems of users' devices.

Considering domestic usage, intelligent management of loads, such as using household appliances, which can be started at any time of the day without particular repercussions on household dynamics, takes place as a result of connection between a smart meter and the network. The meter, in fact, communicates any energy peak time bands upon request of a digital counter using control and information signals. At this point, the smart meter is launched starting the loads only after having confirmed the absence of peaks. In such situations, not only does the peak of the maximum usage flattens out, but it permits to save on the use of stand-by generators or even limits the number of new plants to be built to meet energy needs. Moreover, a considerable economic saving on the part of the user takes place since peak hours are the periods when electricity costs more.

Creation of a smart grid requires greater safety and reliability, given the exposure of such systems to considerable IT problems. Such an architectural infrastructure should be capable of handling almost instantaneous bidirectional communications between each node of the network. Using micro-grids or micro smart grids distributed network architecture would create a much more efficient system than coverage of remote areas.

The problem of electrical coverage of remote areas, such as rural zones and islands, could be resolved using small local power networks of energy supply based on almost total self-sufficiency. Micro-grids are low power networks, which rely on small flow generators. These generators operate as a single system, the purpose of which is to supply electricity to a group of users on a local level.

It is clear that such a network can still manage connections and exchange energy with higher-level networks. Being effective primarily in remote areas, the micro-grid system adopts an "island" operating mode. This means that it will autonomously maintain its generation/load using only local resources such as diesel and hydroelectric generators, and photovoltaic panels. Micro-grids can be distinguished for their ability to manage generators and active loads available in the network with the help of protocols and information flows while coordinating them to perform certain functions in real time.

In this way, smart systems are built based on automated demand management (DSM-Demand Side Management). They are able to disconnect nodes of the network in a selective way according to a modality previously agreed-upon with a user in question (Kupzog F., Tehseen Z., Zaidi A., 2009).

The case study of the present research work called LIDO is inspired by a set of rules proposed by the European Union for smart grids with the intention to introduce a sustainable city model able to optimize distribution and consumption of energy, at the same time reducing its environmental impact.

This set of rules was developed by the JRC (Joint Research Center) of the EU and included 10 guidelines for those who wished to create smart grids and evaluate their practical feasibility according to the classic CBA (cost-benefit analysis) system (European Commission, 2014).

The 10 criteria derived from examining InovGrid project as a case study from which to draw valid conclusions for any smart grid project, at any place and conducted by anyone. The project was led by a Portuguese operator EDP Distribuição and was used as a case study to develop an analysis framework that could be adapted to real-life conditions.

The guidelines formulate various hypotheses according to local conditions, and could be adapted to each context (for example, based on energy demand growth or characteristics of a local network), and in this way allow to identify and monetize costs and benefits in every concrete case that uses a smart grid. The study provides an analysis of critical variables to be considered when working on a smart grid project.

Precisely because it aims to identify if and to what degree realization of a smart grid represents an improvement of the situation existing in a certain context, the guidelines also provide indications on how to identify the so-called "externalities" and social impacts (such as inclusion of consumers, territorial competitiveness). The latter are directly derived from the implementation of smart grid projects, which cannot be easily quantified in monetary terms.

Precisely for this reason, the approach proposed by the EU recognizes that the impact of smart grid projects goes beyond what can be defined in monetary terms and tries to integrate economic analysis (monetary assessment of costs and benefits on behalf of the society) with that of a qualitative impact. The following reference projects were chosen to analyze and ensure connection between systems (smart landscape and smart grid) aimed at creating a network. Various devices and contexts were supposed to bring their identifying characteristics into the structure of the system. In the present research work based on the four smart categories that we consider necessary for creation of a Smart Landscape (S. Grid, S. Community, S. Energy and S. Mobility) we can hypothesize different connections and system contexts that already stand out for their considerable innovation or experimentation.

Smart Landscape derives stability, "nourishment", and can be disseminated and communicated through its main "roots", in particular, through a network of virtual and physical connections that are ensured by the four fundamental categories. Some of the projects are listed below. They were studied first in an exploratory way and later in a more profound manner. This choice is based on the characteristics and potential that they provide for the application of both the concept and the "smart landscape" prototype.

ADDRESS

GRID4EU

ECOGRID

GREEN EMOTION

CRYOGENIC ENERGY STORAGEPILOT

CER SMART METER TRIAL

IMG. 44
REFERENCE NETWORK PROJECT in Europe ill. By Giulia Garbarini

4.3 REFERENCE PROJECT

The following reference projects were chosen to analyze and ensure connection between systems (smart landscape and smart grid) aimed at creating a network. Various devices and contexts were supposed to bring their identifying characteristics into the structure of the system. In the present research work based on the four smart categories that we consider necessary for creation of a Smart Landscape (S. Grid, S. Community, S. Energy and S. Mobility) we can hypothesize different connections and system contexts that already stand out for their considerable innovation or experimentation.

Smart Landscape derives stability, 'nourishment', and can be disseminated and communicated through its main 'roots', in particular, through a network of virtual and physical connections that are ensured by the four fundamental categories. Some of the projects are listed below. They were studied first in an exploratory way and later in a more profound manner. This choice is based on the characteristics and potential that they provide for the application of both the concept and the 'smart landscape' prototype.

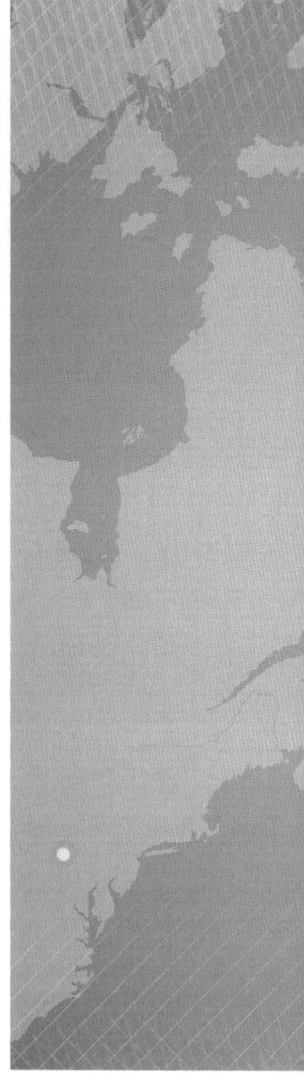

4.3.1 REFERENCE BACKGROUND PROJECT

COPENHAGEN CLIMATE PLAN
CPH 2025 CLIMATE PLAN

ROTTERDAM CLIMATE CHANGE
ADAPTATION STRATEGY

BLUE AP PROJECT

ZEEKRACHT

ENERGY FOREST

SMART GRID GOTLAND

ZWISCHENPALASTNUTZUNG

PLUSNET

ROADMAP 2050

IMG. 45
REFERENCE
BACKGROUND PROJECT
ill. By Giulia Garbarini

4.3.1.1 COPENHAGEN CLIMATE PLAN - CPH 2025 CLIMATE PLAN

Design Team: **Government, organizations and research institutions**
Location: **Copenhagen, Denmark**
Status: **Completed, Future strategy** - Year: **2012**

Characteristics
In 2009, the City Council unanimously adopted 'Climate Plan for Copenhagen', and set a goal: to achieve a 20% reduction in CO_2 emissions by 2015. In addition, a vision for a carbon neutral Copenhagen by 2025 was formulated.

Copenhagen's municipality hoped to reduce CO_2 emissions by additional 20% as its aim was to turn Copenhagen into world's first carbon neutral capital by 2025. At this moment, Copenhagen's municipality is proposing 50 specific initiatives to meet this goal. Those already underway are being appreciated, for example, the initiative which led to achieving a 2014 'World Smart Cities' award for the Copenhagen Connecting project, or the European Green Capital Award. Others are completely new. Some require further preparation or collaboration with state or private sectors, for example, the new PlusNet Strategic Plan, which represents a road planning strategy of slow mobility development.

Furthermore, the City of Copenhagen wants to help create a greener city with a substantial rate of green growth. The Climate Plan will be split into specific goals within four areas: *"energy consumption, energy production, green mobility, and initiatives of the city administration" (Pedersen J. S.)*[60].

With its Climate Adaptation Plan, Copenhagen outlines various challenges a city faces in the short and medium terms because of expected climate changes. It also identifies solutions that, based on our present-day knowledge, appear to be most appropriate and reveal opportunities that climate changes may give to a city.

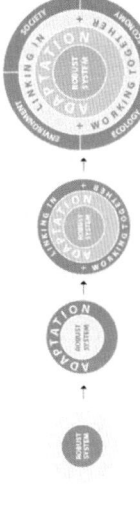

IMG. 46 - 47
Report C40 Good Practice Guides: Rotterdam - Climate Change Adaptation Strategy

60.
See: https://
stateofgreen.com/
en/Profiles/City-of-
Copenhagen
Consulted May, 29, 2017

Potential
The Copenhagen Climate Adaptation Plan not only focuses on minimizing future climate changes but also takes advantage of the adaptation work to improve the quality of life of Copenhageners. For this reason, flexible solutions have to be found with the basis of the existing European policies (Resilient Drivers) plus other pillars (Smart grid, renewable energy, places as catalysts), which will be able to create Smart Landscapes. Urban planning and design should be considered during the work as well as development of recreational areas in the city.

4.3.1.2 ROTTERDAM CLIMATE CHANGE ADAPTATION STRATEGY

Design Team: **Rotterdam Climate Proof**
Location: **Rotterdam, Netherlands** - Client: **Local Municipality**
Status: **Completed, Future strategy** - Year: **2007**

IMG. 48
Copenhagen 2025 full report

Characteristics
Rotterdam has an integrated climate change adaptation approach, marked by adoption of the Rotterdam Climate Proof (2008) and the Rotterdam Climate Change Adaptation Strategy (2013).
The strategy aims at:
a) strengthening a robust system of flood control; b) adapting urban space to combine its three functions: 'sponge' (water squares, infiltration zones and green spaces), protection and damage control; c) increasing the city's resilience through integrated planning; d) fostering the opportunities brought by climate change, such as strengthening the economy, improving the quality of life, and increasing biodiversity.
Rotterdam's adaptation system is based on the flood and sea-level rise defence system, consisting of Maeslantkering (flexible storm surge barrier), permanent sand dunes along the coast, and dykes along the rivers. In addition, Rotterdam

is also addressing heavy rainfall threats. It has built water storage spaces, including the Museumpark car park underground water storage with capacity of 10,000 m3, and is integrating Blue-Green Corridors into the urban landscape. Rotterdam has also installed over 185,000 m3 of green roofs in 2014 alone. Finally, the launch of a 100% climate-proof neighbourhood-scale project in the Zomerhofkwartier, demonstrates Rotterdam's commitment to comprehensive delta city adaptation.

Potential
In this context, the Rotterdam Climate Change Adaptation Strategy offers many opportunities to strengthen the economy of the city and port, to improve the quality of life in neighbourhoods and districts, to increase biodiversity in the city and to foster committed and active participation of Rotterdam's residents in the city's life. Working with the concepts underlying the procedures of European policy for a climate-proof city is a perfect common ground for application of an energy smart grid model that connects pilot projects and innovations within climate adaptation measures but not only. Combining the existing European policies (Resilient Drivers) with other pillars (Smart grid, renewable energy, and places as catalysts) will create a 4.0 landscape as network, a landscape that will be able to become "smart".

4.3.1.3   BLUE AP PROJECT

Design Team: **Coordinator Giovanni Fini**
Location: **Bologna, Italy**
Status: **Completed, Future strategy** - Year: 2012

Characteristics
BLUE AP (Bologna Local Urban Environment Adaptation Plan for a Resilient City) is a LIFE+ project of implementation of Climate Change Adaptation Plan by the municipality of Bologna. It provided some concrete local measures to be tested in order to make the city more resilient and able to meet climate change challenges.

IMG. 49
BlueAP (Bologna Local Urban Environment Adaptation Plan for a Resilient City).
Project partners: Municipality of Bologna, Kyoto Club, Arpa Emilia-Romagna, Ambiente Italia.
Graphic by: TAKK

The Adaptation Plan contains measures that help to deal with potential drought and water shortages, urban heat waves, and excessive rain and hydrogeological risks. It also identified a series of 'green' and 'blue' best practices that could be implemented in Bologna and other Italian cities. The 'green' measures are: periurban parks, green and cool roofs/walls etc.; the 'blue' measures are: permeable pavements, rainwater collecting etc.

The Plan outlines strategies that confront critical situations highlighted in the POC and identifies a series of good practice actions with the objective of completing all of them by 2025. The Plan regards not only actions themselves, but also the way of implementing them, and pays particular attention to interaction between different levels of local government and individuals. The socio-environmental goal is to prepare institutions and citizens to face the consequences of climate changes more effectively, while reducing existing vulnerabilities in the Bologna area.

Potential

The BLUE AP project shows the efficiency of LIFE as a tool for sub-national and municipal adaptation planning. However, it is important to synchronize the impacts of bottom-up initiatives with the top-down imperatives of national adaptation strategies and ensure strong connections between sectors. The following processes highlight the possibility of developing a national strategy that can be effectively implemented at a local level, starting with encouraging urban renewal aimed at reduction of soil consumption and regeneration of areas with abandoned buildings (place as catalyst). Combining the existing European policies (Resilient Drivers) with other pillars (smart grid, renewable energy, and places as catalysts) will create a 4.0 landscape as network, a landscape that will be able to become "smart".

4.3.1.4 ZEEKRACHT

Design Team: **Oma**
Location: **North Sea, Netherlands**
Client: **Natuur en Milieu**
Status: **Completed, Masterplan study** - Year: **2008**

ZEEKRACHT

Characteristics

OMA (Office for Metropolitan Architecture) has recently presented a masterplan for the North Sea, claiming that wind farms in the North Sea could produce as much energy as the Persian Gulf oil is producing now. If built, the giant Zeekracht wind facility could give Europe energy independence. Zeekracht, a masterplan for the North Sea, maps out a massive renewable energy infrastructure engaging all its surrounding countries (and potentially those beyond) in a supranational effort that will be both immediately exploitable and conducive to decades of coordinated development. Primary components of the Zeekracht masterplan include an Energy Super-Ring of offshore wind farms – the main infrastructure for energy supply, efficient distribution, and strategic growth; the Production Belt – the on-land industrial and institutional infrastructure supporting manufacturing and research; The Reefs – integrating ecology and industry by stimulating existing marine life alongside wind turbines and other installations; and an International Research Centre – promoting cooperation, innovation and shared scientific development.

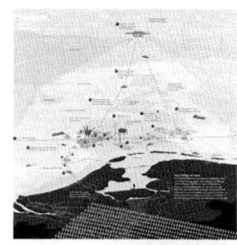

Potential

The Zeekracht is conceived as a reciprocal system, fed and reinforced from the top down in terms of technology, industrial development, and Europe-wide policy; and from the bottom up in terms of local decision-making, popular involvement and support.

Unlike the usual planning methods based on least-conflict zoning, the masterplan suggests a multi-dimensional approach based on optimizing potential. Productivity and profitability of offshore wind farms can be enhanced if they synthesize with activities existing in North Sea (such as shipping and oil and gas extraction) and new programs (such as eco stimulation

IMG. 50 - 51 - 52
Zeekracht - A Strategy for Masterplanning the North Sea. 2008.
By OMA studio.
Partner: Rem Koolhaas, Reinier de Graaf
Project director: Art Zaaijer

and tourism). By sharing and combining information acquired for this masterplan and information received from our models of processes, these technologies could go far beyond today's standards and create a renewable energy network based on smart grid structure, which is related to European policies and to places as tourism catalysts.

4.3.1.5 ENERGY FOREST

Design Team: **Stoss - Collaborators: MY Studio/Höweler + Yoon Architecture, Nitsch Engineering, LightTHIS!**
Location: **Pittsburgh, PA, USA - Client: Sports and Exhibition Authority**
Status: **Completed (Competition) - Year: 2009**

IMG. 53 - 54 - 55
Energy Forest, diagram and vision.
Project by: STOSS studio

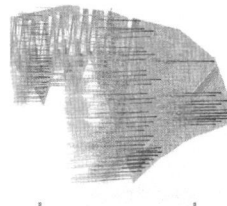
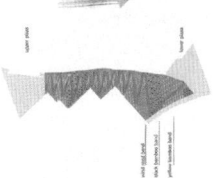

Characteristics

Energy Forest is a new type of urban plaza, a dynamic and sustainable civic grove. The forest that was designed for this urban plaza consists of alternating bands of black and yellow bamboo and steel energy poles, which accentuate the journey from down to up, or up to down.

The project's technological elements such as vertical metal poles in the forest and illuminated water walls at the embedded end of the pathway create a new set of ecological tools that are both sustainable and productive. The poles capture and turn in wind energy, creating intermittent lighting effects in the forest. In contrast, low water walls on the upside of the switch-back utilize available rainwater to illuminate a matrix of led fixtures planted into its surface. Together, the poles and walls pulse alternately and simultaneously in a constantly evolving light play, bringing a sense of life and energy to the hillside.

In this STOSS project "Energy Forest" plaza in Pittsburgh, we see designers deploying rhetorical power of the grove, and the forest to provide a deeper reading to engineered infrastructural or formal constructions.

Potential

The 'Energy Forest' project is a case study, which makes it clear that strategies and technological models designed to address future questions from the point of view of European landscape are changing into those of smart landscape. This

Stoss project would be capable of integrating into urban or rural energy policies with the help of urban, landscape and social devices. It would be possible to use these devices as enabling instruments for a different landscape level. Application of requirements and provisions of this process creates a schedule that not only offers technical benefits but also turns cities into more sustainable and attractive. Smart grids could deliver reliable and efficient infrastructure and facilitate exploiting a model of interconnected networks that will become a track for creation of places as catalysts of values. This process will change the smart grid model with the ambition to unite other compatible and enabling technologies and the European policy.

4.3.1.6 SMART GRID GOTLAND

Design Team: **GEAB, ABB, Schneider Electric and Echelon**
Location: **Island of Gotland, Sweden**
Client: **National and Local Municipality**
Status: **In progress** - Year: **2012**

Characteristics

Smart Grid Gotland is a project that aims to upgrade the existing rural distribution system of a deregulated market to a modern smart grid. The Smart Grid Gotland R&D project intends to develop strategies for planning, construction and operation of a fully developed, large-scale Smart Grid, including a large scale of intermittent production, primarily from wind power, in a distribution network. *"New market models and services will be developed to involve active customer participation and pave the way for new market players"*[61].

More specifically, the existing distribution network will be turned into a smart grid to deal with challenges associated with an increased proportion of renewable electricity generation. The integration of Wind Power with existing distribution network ensures a reliability and efficiency. It shows that modern technology can help increase power

61.
Energy Policies of IEA Countries Sweden 2013 Review.
Available on:
http://www.iea.org/publications/freepublications/publication/Sweden2013_free.pdf

quality in rural networks with large amounts of installed renewable generation.

With this development of the future smart distribution grid, consumers and producers will be fully integrated in project that is likely to become an international model for a long-term sustainable electricity power system.

Potential

The Smart Grid Gotland R&D project intends to develop strategies for planning, construction and operation of a fully developed, large-scale Smart Grid including a large scale of intermittent production, primarily from wind power in the distribution network. This pilot case also provides a step towards full-scale implementation of Smart Grids in Sweden. However, will it be possible to use the structure of the smart grid as an architectural driver for building a smart landscape?

The potential that resides in the Smart Grid Gotland can be compared with a complete power system of an energy smart grid that includes various elements. It has the possibility to turn it into an architectural driver, which will make it possible to build a new process (Smart Grid as Architectural Driver).

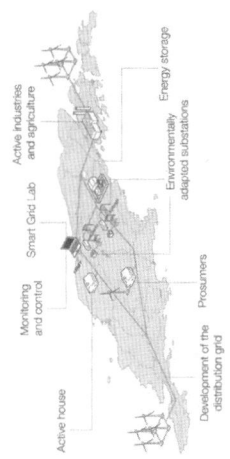

IMG. 56 - 57
Project diagrams "Smart grid gotland".

4.3.1.7
ZWISCHENPALASTNUTZUNG

Design Team: **Philipp Misselwitz, Philipp Oswalt, Klaus Overmeyer**
Location: **Berlin, Germany**
Client: **International Commission**
Status: **Completed** - Year: **2002**

Characteristics

After being used for fourteen years, the Palast der Republik was closed in 1990 for asbestos abatement. Conducted from 1998 to 2001, it reduced the building to a skeleton. Politicians were about to demolish this important architectural symbol of communist East Germany, but the historical site and fascinating space inside aroused interest in the cultural sphere. Thus in Summer 2002, together with a group of those interested in using the building, a group called Urban Catalyst developed a realization plan, which included a spatial concept, expert planning for traffic safety, fire protection, building costs, and a financing plan. Among partners were the Berlin State Opera, Sophiensäle, Museum of Transport and Technology, WMF Club, and the artist Fred Rubin. In November 2002, the project was presented to the public at the exhibition in the Staatsratsgebäude Berlin (former East German parliament building). Public resonance was very strong and positive, but the owner (German federal government) initially stopped those interested, claimed that the building was not safe for usage, and refused constructive dialogues. Only through a complicated process lasting more than two years and involving systematic realization of small-scale activities was the owner finally persuaded of the possibility and opportunity to use the building for cultural purposes. Starting from spring 2004, various parties had rented the Palast der Republik. By December 2005, 916 events had drawn more than 600,000 participants. The central event was Projekt Volkspalast ('People's Palace'), a festival lasting several weeks in 2004, and again in 2005 with a spectrum of events in a very wide range of areas attracting international attention.

Potential

The potential of this project managed by the collective Urban Catalyst resides in the design techniques. Urban Catalyst conducted a research, in which different intervention and design strategies were implemented to create a temporary effect. It revealed several examples of both success and failure that have been realized in cities such as Berlin, London, Vienna, Rome and Amsterdam.
The project in question was inserted into the Berlin context where a designer is no longer the one who takes decisions, but the one who activates and connects other participants of the initiative. Users thus become creators of space.
This strategy, as well as the potential that this project contents, is called CLAIM, that is: 'creating new public spaces that generate new social and cultural impulses'. The goal, and therefore its potential, was to create a cultural space while preserving the historical Palast der Republik that had stayed closed for 12 years.

IMG. 58
Volkspalast
From August, 20 to November, 9, the Palace of the Republic will be used for cultural purposes.
By : Urban Catalyst group.

This made it possible to launch numerous cultural activities including a space for performances and a theater.

The project was initially rejected by the government because it was considered too complicated, but later it was accepted by the administration, which carried out the work of securing the building. This fact demonstrates that a cultural action can activate even the most rigid administrations.

Characteristic

Copenhagen represents a city model, which unites the importance of sustainable mobility and urban planning. This peculiarity might be interesting in the context of transferring it to the Italian reality.

The cycling infrastructure plan is partly based on a survey covering some issues and proposals, information on existing cycling patterns, siting of primary corridors/links, and major cycling destinations (workplaces, institutions, shops). All initiatives of this kind should be based on a bicycle infrastructure plan, sometimes referred to as a "cycle track plan", or should be integral part of a "cycling action plan", which includes many different aspects of cycling promotion. Slow-speed zones and secondary roads with little motor traffic may be included in the cycling area.

The objective of the Plan is to create a network of capillary and continuous cycle paths throughout the urban territory until reaching the extra-urban area adjacent to the city.

IMG. 59 - 60
ZWISCHENPALASTNUTZUNG
Photo by: Urban Catalyst group.

5.3.1.8 STRATEGIC PLAN PLUSNET

Design Team: Government, organizations and research institutions
Location: Copenhagen, Denmark
Status: Completed, Future strategy
Year: started in 2011

Currently, Copenhagen offers approximately 400 km of cycle paths, two bridges left to bikes only, and a safety system extended to the entire cycle network. As part of the PlusNet project, the types of tracks to be built are substantially two (three lanes, when the road is one way, and four lanes, when there is a double direction of travel), which became a key point of the project.
Within the PlusNet project exist:
1. City life: re-planning, by 2025, of all existing city arteries in order to change the point of view on mobility;
2. Comfort: integration, by 2025, of the new cycle paths with existing ones. New parking lots are planned to coincide with the areas with high commercial density and near the 'business' areas. There are also new bike sharing systems integrated with the city transport network for a fast and problem-free 'bike parking' in the metropolitan area;
3. Travel Time: rationalization of slow mobility travel time as a result of construction of new infrastructures, such as bridges and cycle paths in green areas of the city, and of the expansion of the already constructed slopes;
4. Sense of Security: PlusNet wants to increase the sense of safety of cyclists and that is why it projects widening of the roadways and improvement of horizontal road signs.

Potential
PLUSnet ensures high quality of space, intersections, and maintenance so that numerous cyclists can travel securely and comfortably at pace that suits each individual. On the PLUSnet, a Copenhagener can converse with a friend or cycle next to their family without being disturbed by the bellsof people who want to get past, as there are 3 lanes in each direction on

IMG. 61
Plan of the project PLUS net.
By: Government, organizations and research institutions

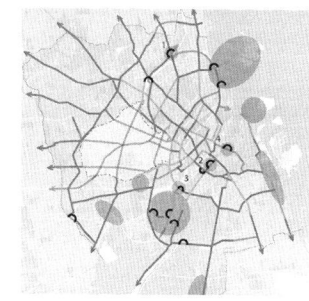

80% of the network (4 lanes in total on stretches where the cycle tracks are bi-directional). The map shows examples of large-scale improvements that have been approved and others with high priority to be made by 2025. The exact routes and capacity will be adjusted on an ongoing basis depending on traffic and city's development projects (The city of Copenhagen's Bicycle Strategy 2011-2025, The Technical Administration Traffic Department). Cargo bicycles, for example, are not so common in Italy, while in Copenhagen they are used for transporting children and grocery, and are often an alternative to car. Actually, one-fourth of all cargo bicycle owners say that it is a direct replacement for a car. By 2025, excellent parking facilities for cargo bicycles are going to be created outside of homes, institutions, and shops. In order to encourage more people to use bicycles, it is important to establish travel times that are competitive with the other transport forms. For this reason, the project focuses mainly on various strategic plans that have been launched in the Danish capital, from cycling tradition to the first strategic plan that aims at creating the Vision 2025.

5.3.1.10 ROADMAP 2050: A PRACTICAL GUIDE TO A PROSPEROUS, LOW-CARBON EUROPE

Design Team: Imperial College London, KEMA, McKinsey & Company, Oxford Economics and AMO (the think tank within the Office for Metropolitan Architecture (OMA)).
Client: European Climate Foundation
Status: Completed - Year: 2010

Characteristics
The 'Roadmap 2050' research project was commissioned by the European Climate Foundation, a philanthropic body dedicated to promoting policies that reduce greenhouse gas emissions. It aimed at showing how the EU could achieve a

seemingly unreachable target of an 80% reduction in carbon emissions by 2050. The proposal was accepted by the EU Council of Ministers.

The proposal's starting point is the fact that renewable energy sources such as wind and sunshine are erratic and unreliable, which means they have to be supported by other forms of power. However, they are also available in different quantities in different places – wind is abundant in Britain, sun in Spain – and in different seasons. The main idea is to create a power network across the continent linking all these sources, which could then compensate for each other. If it were windless in Britain but sunny in Spain, power could be transferred from one to another, and vice versa.

This is a political and technical proposal, which coincides with the work the OMA office has been conducting for several years. According to the 'roadmap 2050', 80% CO^2 reduction overall implies 90-95% reduction in power, road transport, and construction sectors. This could be achieved by maximum abatement within and across sectors. The most influential sector will be power and vehicle transportation. The level of decarbonization is dependent on achieving aggressive 2% energy efficiency savings a year, without which this level of abatement is not possible in this model.

The project is based on European leaders' commitment to an 80-95% reduction in CO^2 emissions by 2050. The technical and economic analyses outline why a zero-carbon power sector is required to meet this commitment and illustrate its feasibility by 2050 given current technology progress. The project then aims to chart a policy roadmap for the next 5-10 years based on the near-term implications of this commitment. Through complete integration and synchronization of the EU's energy infrastructure, Europe can take maximum advantage of its geographical diversity. The results of the report show that by 2050, simultaneous presence of various renewable energy sources within the EU can create a complementary system of energy provision ensuring energy security for future generations.

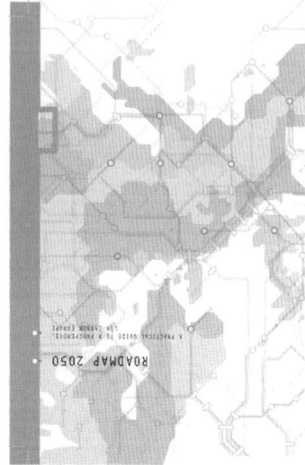

IMG. 62
Report of: "ROADMAP 2050: A PRACTICAL GUIDE TO A PROSPEROUS, LOW-CARBON EUROPE."
Edited by OMA/AMO

IMG. 63
Diagram of the ROADMAP 2050: A PRACTICAL GUIDE TO A PROSPEROUS, LOW-CARBON EUROPE.
Edited by OMA/AMO studio

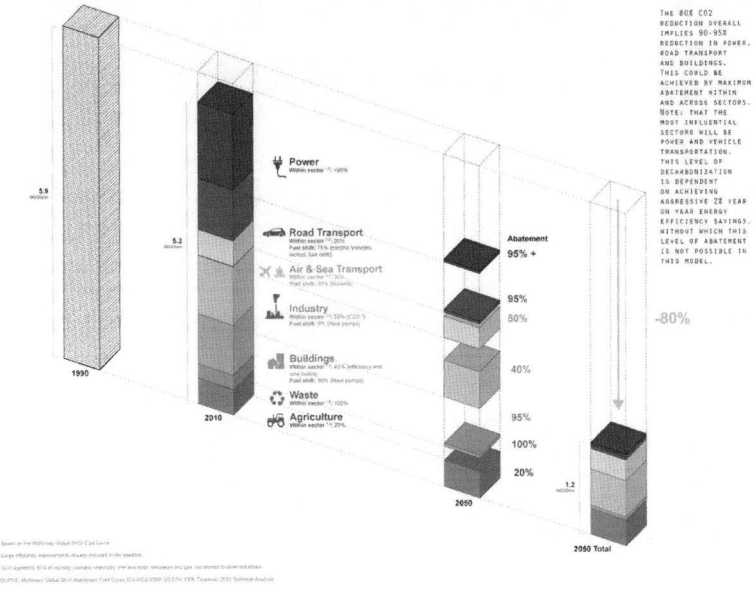

IMG. 64
Diagram of the ROADMAP 2050: A PRACTICAL GUIDE TO A PROSPEROUS, LOW-CARBON EUROPE. Edited by OMA/AMO studio

Potential

This project focuses mainly on the vision. This is one of the most avant-garde research projects / proposals that are based on the aim of architecture as science to respond to the future problems related to climate change.

It will project Europe into a futuristic vision of cohesion and connection, of energy and values, capable of facing the same problems of climate change in progress.

The potential of the project is contained in the same description that is made by OMA's partner in charge of the project, who said: "in our profession there is a lot of talk about sustainability, but it is generally only dealt with at the scale of buildings. This project allowed us to address the issue at an entirely different scale. In the end, the planning of a trans-national renewable energy grid has a much larger impact and more widely shared benefits." The project is based on two foundational AMO projects: European exhibitions, commissioned by the Dutch EU presidency in 2004 (which included the proposal for a composite 'barcode' EU flag), and Zeekracht, a 2008 plan of building a ring of offshore wind farms in the North sea.

4.3.2 REFERENCE MAIN PROJECT

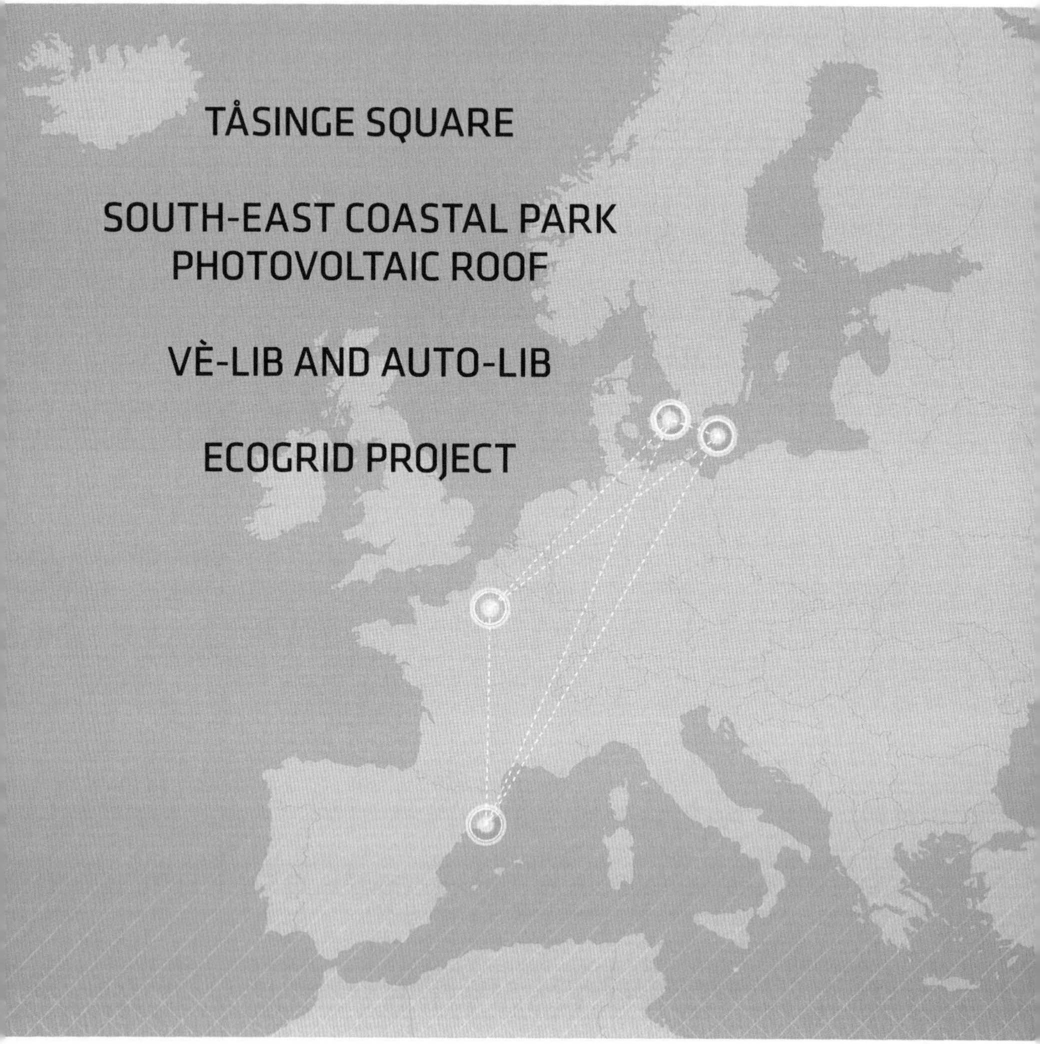

TÅSINGE SQUARE

SOUTH-EAST COASTAL PARK
PHOTOVOLTAIC ROOF

VÈ-LIB AND AUTO-LIB

ECOGRID PROJECT

REFERENCE
MAIN PROJECT
ill. By Giulia Garbarini

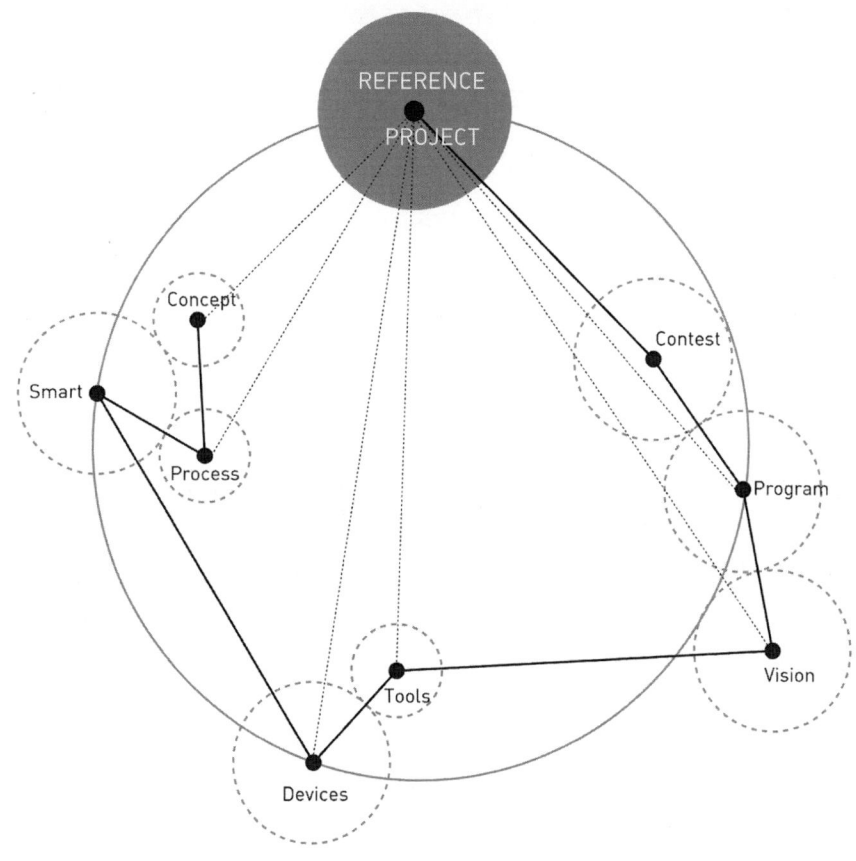

IMG. 65
Reading diagram of the main project.
img. By Giulia Garbarini

4.3.2.1 TASINGE SQUARE. COPENHAGEN
55°42'36.1"N 12°34'04.4"E

Design: **GHB Landskabsarkitekter A/S**
Entrepreneur: **Malmos Anlaegsgartnere**
Engineer: **Orbicon**
Consultants: **Feld Studio for Digital Crafts**
Client: **Copenhagen municipality at Center for Park and Nature,**
Area renewal Skt. Kjelds District, Copenhagen Patios, Copenhagens energu and Environment Østerbro.
State: **Completed**
Year: **2013-2014**
Size: **7.500 m²**

CONTEST

The same year Tåsinge Square was completed, the city of Copenhagen received the World Smart Cities Award at the Smart City Expo World Congress in Barcelona.

Today, Copenhagen is still considered to be one of the cities with the best living standards in the world, and is a holder of various awards, ex. "European Green Capital" of 2014. Copenhagen is certainly one of the most complete smart cities currently existing, and it is aiming to become a Carbon Neutral City by 2025. The capital of Denmark has established numerous and ambitious goals in the field of energy efficiency, the use of renewable energies, green building standards, in the bland mobility and, last but not least (and especially important for the present research work), it is expanding an integrated system for managing and distributing energy through a single, more dynamic and controllable system, Smart Grid (Nyborg S., Røpke I., 2013).

The city administration has launched 47 projects as part of the currently in progress "Carbon Neutral City 2025". Some of the various programs include: private partner investments for an estimated amount between DKK 200 and 250 billion (between $34,800 and 43,500 million); public investments up to 2025 of DKK 2.7 billion (about $470 million). This funding has been allocated to respond to the needs and problems related to the era of globalization and, in particular, to water. In fact, one of the peculiarities of the city of Copenhagen is its close connection to water, for which it is often called the Venice of the North, and, like for the Italian city, water for Copenhagen is r a fascinating feature, though it can also involve several issues. However, it is more interesting to note how the city gathers strength for new measures to cope with its weaknesses, and, as mentioned in the conference "Cities by Water: Solutions from Copenhagen and New York" held on August 4, 2014 at AIA, New York, *"The question is not whether to act, but how to act. How do we increase the resiliency and sustainability of our coastal cities?"* (Mondkar B., 2014).

Inside the project "Carbon Neutral City 2025", after years of planning its climate adaptation effects, Copenhagen is ready to launch the first 16 projects (Technical and Environmental Affairs of Copenhagen, 2015). This will become an additional catalyst of changes for the city, which will have to evolve fast to make climate of Copenhagen resilient.

One of the projects carried out (and of much interest for this thesis) is located in the Østerbro district, which makes part of the Copenhagen Climate-Resilient Neighborhood. This project addresses the issue that the city of Copenhagen, like many other cities in Europe, has always confronted: the future increase in amount of rainwater.

> *"The increase in rainfall is a major challenge for our city. But by tackling the challenge the right way, we can secure the city from cloudbursts while also bringing the city new recreational values. The ideas in Saint Kjeld's Neighborhood are a really good example of this,"* says Technical and Environmental Mayor of Copenhagen, Ayfer Baykal (SF)" (Furuto A., 2012).

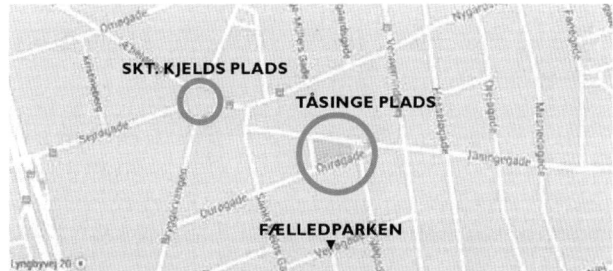

IMG. 66
Localization TASINGE SQUARE.
All illustrations by TREDJE NATUR

PROGRAM

The project of Tåsinge Square, as explained by Alison Furuto (2012) in her article, is a part of the plan to develop Denmark's first climate adapted neighborhood, which transforms Skt. Kjelds District, outer Østerbro, into Copenhagen's greenest area. This comprehensive urban development project seeks to demonstrate how the city can be arranged so that rainwater on the streets can be managed in a more natural and effective way. Østerbro is a particular neighborhood, where changes were to be made with the idea of general consciousness about climate change and broader questions like "what kind of world would we like to give to our children?" Here. it was necessary to act in a conscious manner in order to show what individuals can do to make the world better through changes in urban spaces, and ensure the neighborhoods of Copenhagen their position of influence in the society by reaching ambitious goals of the 'Climate Neighborhood'.

It was important to activate the city in a sensory way and to create an ethical framework that produces scenarios in which people can observe functional and characteristic changes of the city, which could lead to reflections on these changes. It is therefore critical that designers work with social and sensory optimization to make people aware of consequences of their actions.

As the Copenhagen Climate-Resilient Neighborhood report illustrates (Klimakvarter, 2015), transformation of Tåsinge Square was one of the first projects realized and completed with respect to the goals of the 'Carbon Neutral City 2025'. It is just one of the numerous projects related to St. Kjeld and linked to the Climate Adaptation Plan. In fact, the square in question makes part of a wider intervention plan that is to be adapted in the neighborhood. It consists of a mixed social component, which contains several ethnicities, and it was tried and developed on a test area aimed at becoming the first and most resilient and adaptive neighborhood in the optics of climate change.

This program is designed not only as technological development but also as a social project.

The "*project started in 2012, focusing on St. Kjeld's Square, Tåsinge Square and Bryggervangen*" *(kllmakvarter.dk)*. The goal was to finally re-open the squares in 2015/2016 to show people and administrations that climate adaptation solutions could serve as a source of inspiration in creation of greener streets and improved urban spaces throughout various cities around the world. The project contains many layers of interest and application, including: light, entrances, traffic, cables and pipes, terrain. (Klimakvarter, 2015).

Climate adaptation projects are a great opportunity to add value to cities. However, it takes continuous political commitment to allocate the necessary funds. The public square in Eastern Copenhagen, Tåsinge Square, was a perfect illustration of why many major cities in Europe and in the world struggle to adapt to climate changes.

> *"Most cities are now realizing that covering huge areas in asphalt and pavement is a certain way to create floods during extreme rain events. The water has nowhere to go but into the sewers or remain on the surface. Unless a city has very deep pockets and can build huge sewer pipes underground, the most cost and active way to manage precipitation is by catching and delaying the water on the surface, and subsequently directing it in a controlled way into underground pipes..."*

says Henriette Berggreen of the City of Copenhagen. Copenhagen is working diligently to make this plan work. The city has laid out a program to adapt the capital to the effects of climate change. Tåsinge Square, along with the entire neighborhood of St. Kjeld's, is pioneering the transition (Technical and Environmental Affairs of Copenhagen, 2015).

The whole area counts potentially 50,000 m^2 and is to be used for new urban space development, such as Saint Kjeld's Square and Tåsinge Square, where the project has created new kinds of urban experiences founded in the city's and nature's changeability.

> *"At Saint Kjeld's Square and Tåsinge Square we optimize the terrain effectively doubling the urban space's surface area. This provides new space for a comprehensive volume of natural value, a better micro-climate and hence more urban life and better rain water management. The large spaces will be both striking and natural hot spots in the district. It is also possible to integrate new cafes and playgrounds in the optimized terrain,"* says Flemming Rafn Thomsen, Partner in Tredje Natur (Furuto A. 2012).

VISION

The first "vision" of the project comes from the winners of the "Europan 11" international competition[62]. In that edition, the project conducted by Tredje Natur studio inside St. Kjelds Neighborhood, won the first prize and let the studio expand the proposal and widen it.

Named "The First Climate District", it involves the whole neighborhood of Østerbro and demonstrates that adaptation to climate change can become a distinctive feature of Copenhagen's city plane

The square of Tåsinge, designed by GHB Landskabsarkitekter, is collated within the above mentioned project. Today, Tåsinge Square looks very different from how it used to be. The square has been turned into an abundant green oasis in the middle of the capital. Moreover, an innovative hydraulic and social purpose of the redesign was to detain and redirect as much of the precipitation that falls to the square.

"Securing a city like Copenhagen against climate change is a huge investment, and we need to make sure that this investment benefits our citizens' everyday lives," explains Henriette Berggreen.

Therefore, Tåsinge Square today is a gathering point for the entire neighborhood; it is similar to a "smart" park, which uses local weaknesses as a future driving force where the final design of the square was decided with the help of the area's residents. Even this combination of technological attention to the best adaptation strategy and social attention to make this place important for the citizens denotes a "smart" feature. The square is situated on a slope and will be an important element in a network of "Cloudburst Roads", roads designed

62.
IMG. 67
Vision of the "Europan 11" international competition by Tredje Natur studio inside St. Kjelds Neighborhood.
img. by : Tredje Natur studio

to contain and direct rainwater to parks, reservoirs or harbors (Technical and Environmental Affairs of Copenhagen, 2015).
Tåsinge Square *"is an urban habitat in which the city's rhythm meets nature's cycles and the logic of rainwater*

IMG. 68
The square of Tåsinge.
Photo by: Tredje Natur studio

*forms urban environment. The activity of citizens, water flows, precise geometry in the district and Copenhagen dialect are all combined with the natural and self-grown approach to vegetation and water.
The topography creates an urban space where edges, transitions, and the relationship between inside and outside are essential"* (GHB Landskabsarkitekter a/s, 2014).

Copenhagen experienced a short but intense cloudburst during the summer of 2015, which put the climate adaption project to the test - the test that was passed with success. Rainwater from the streets was collected efficiently through holes in the kerbs and led to the mould water beds, while rainwater from the nearby roofs was directed to the underground reservoir. This ensured that the neighborhood's buildings and basements avoided flooding and damages.

DEVICES AND TOOLS
Referring to the brochure outlined by Klimakvarter.dk (2015) and titled "TÅSINGE PLADS - En lokal grøn oase, hvor regnvand skaber rammer for leg, ophold og nye møder", Tåsinge Square is a very peculiar public space. Within this research work it is not possible to outline one single device that enables

IMG. 69
The square of Tåsinge.
Photo by: Tredje Natur studio

the context to be smart, but a series of related actions, devices and tools that, put together, lead to the creation of a comfortable and functional common space. The project combines successful engineering and electronic devices, which are capable of meeting Copenhagen's objectives that regard climate change. However, at the same time, they are able to become a reference for the neighborhood and for those who inhabit it. This goal, already reached today, was achieved thanks to a strong dialogue with the inhabitants of the district and thanks to a process characterized by projecting and realization of both small and massive interventions of urban renewal and innovative installations, which involved citizens in the choice process[63]. The same spatial layout of the square was conceived, as can be seen in section[64], within a "classic and familiar" Danish landscape (Klimakvarter.dk 2015), which extends from rolling hills to the shores of a lake. Based on this concept we can observe how the square starts with a slight slope to the east, where water from the roads is collected in so-called "water beds", in which some plants are placed. These plants are able to resist for days having their roots immersed in water, thus creating

63.

Borgerdrevet. By fornyelse_Bølgen. Urban renewal called in English "The Wave", which was conceived for a contest of ideas in 2013 and was subsequently accepted by neighborhood residents as temporary renewal inside the square.

a remarkable biodiversity of the square. In fact, an aspect of equal importance to the climate one, in the process of redesigning the square, was that of demonstrating how the flora used in the present context is able to improve the quality of life of neighborhood's inhabitants from a social point of view and not only as a factor of atmospheric agents' mitigation. (Technical and Environmental Affairs of Copenhagen, 2015). But the main objective remains water management.

In fact, this problem is addressed and managed with the use of three different devices, which can be defined and identified as enabling technologies[65], especially if we think of a way they rend resilient a space that previously suffered from floods.

First of all, in the center of the square we find "black umbrellas", which have the function not only to protect pedestrians from rain, but also to channel the rainwater and to let it concentrate in the underlying tanks without disturbing the passers-by. Another feature characteristic operating in the opposite way are the so-called "drops of water"[66], which recall a drop of water that falls to the ground, and, just as in reality, inside their transparent casing the drops contain water introduced with the help of a pump (preferably during the summer months). This system leads to creation of a play area for children of the neighborhood. The water that is first pumped inside the "drop" and then flows out comes from other underground tanks, where the water from the drainage channels of the houses surrounding the square is conveyed, filtered and

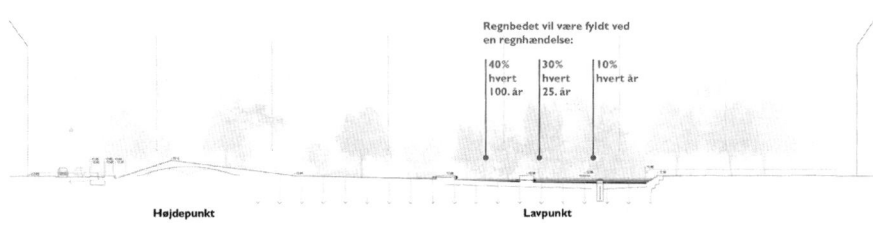

64.
IMG. 70
Section of the square project of Tåsinge.
img. by: Tredje Natur studio

stored. The above-mentioned excess water, which is channeled by the systems illustrated above will then be conveyed to the municipal sewage system (GHB Landskabsarkitekter 2014; Klimakvarter.dk 2015).

Water on the road pavement is collected and conveyed into so-called water beds, provided with special flooring that allows channeled water to be filtered before reaching underground tanks. This system focuses on the climate adaptation of the district and at the same time creates a new topography of the square, which is able to integrate an innovative conformation of the space, promoting technological, environmental and social interaction in a unique, new and peculiar use of Tåsinge Square (GHB Landskabsarkitekter, 2014).

SMART PROCESS

In this case, the process that is being implemented in this context does not only concern the application of so called enabling technologies or ICTs. The latter can be adapted to the Klimakvarter pilot project, which is part of the Copenhagen's adaptation plan intended to provide solutions to the three main types of impact: rising precipitation level, rising sea level and heat waves. In this case, the process can be defined as "smart" as it takes a continuous and variable measures based on changing needs. It is using flexible solutions that can increase the quality of life and generate positive economic outcomes. All the features and purposes are essential to the

65.

IMG. 71
Depicting the enabling technologies used.
img. by: Tredje Natur studio

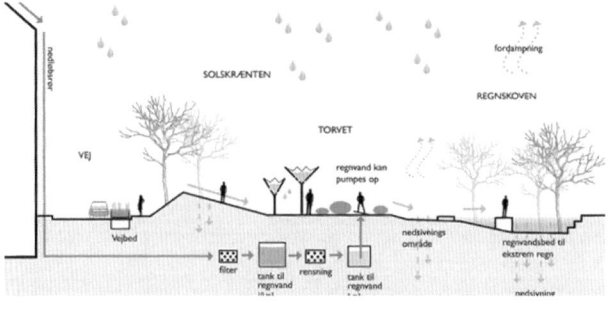

66.

IMM. 72
Drops of water concept and vision
img. by: Tredje Natur studio

selection of priorities to be adopted.
Such an approach and combination of technologies used make this intervention an insight into the universe of the Smart Community actions/processes. To define it we use the description by Eger:

> *"Smart community – a community which makes a conscious decision to aggressively deploy technology as a catalyst to solving its social and business needs – will undoubtedly focus on building its high-speed broadband infrastructures, but the real opportunity is in rebuilding and renewing a sense of place, and in the process a sense of civic pride. [. . .] Smart communities are not, at their core, exercises in the deployment and use of technology, but in the promotion of economic development, job growth, and an increased quality of life. In other words, technological propagation of smart communities isn't an end in itself, but only a means to reinventing cities for a new economy and society with clear and compelling community benefit." (Eger J.M., 2009)*

Communities around the world are responding to the needs of their citizens by discovering new ways of using information and communication technologies for economic, social and cultural development. Companies and governments that take advantage of these new technologies will create jobs and economic growth as well as improve the overall quality of life within the communities (Lindskog H., 2004).
In this particular project, the components and the process of ideation and construction are aimed at safeguarding inhabitants of this place and at revaluation of its original weaknesses. The implemented technologies are part of a wider community plan, which aims at expanding throughout the city, thus creating, as it has already been the case in this place, numerous urban catalysts. Various areas could become catalysts thanks to ensuring security resulting from the use of appropriate devices dislocated around the area and progressive participation of the inhabitants. The community wants to consciously increase the usability of spaces and thinks and targets technology as the only possible tool

for resolving corporate and social issues. But the real opportunity must be taken from reconstruction and renewal of a sense of a place and a process of civic pride, as in the present case.

SMART CONCEPT

Tåsinge Square is the first Copenhagen-based urban climate-adapted area. The square (7500 m²) is a green oasis, which creates meeting places for residents of the neighborhood, where it was possible to condition and manage the water flowing into the square and coming from the roofs.

Within this project, as it should be in other urban contexts, the diffusion of enabling technologies becomes a means of reinventing the city for the benefit of the community.

In a smart city such as Copenhagen this fact becomes evident through various initiatives, but it is within the project of Tåsinge Square that we can notice a strong desire of the inhabitants to promote and participate in the growth of their neighborhood and the city, as this desire is the basis of the whole project.

Based on these considerations, the project, like the plan, within which it developed, adopts a green approach and proposes a project for the development of "green" and "blue" measures to be coordinated with the project of implementation of water collection system in addition to the strategy for new plantings in green areas, parks and greater use of the square by citizens.

This solution is adopted as expansion of the so-called "gray" infrastructure (i.e. management of rainwater and usual water infrastructure) and is judged and estimated to be too costly, excessively long to be achieved and having results of dubious effectiveness.

It is left out to favor application of "green" solutions at a local level, thus leading to the redesign of road sediments and introduction of green spaces that contribute to the renewal of urban structure and at the same time redirect water away from residences.

As illustrated by Filpa and Pellegrini (Filpa A., Pellegrini V., 2013), this redesign has consistently involved road surface resulting in the modification of 20% of the asphalt surfaces into green areas, aiming at retaining 30% of rainwater in the subsoil. The reduction in the road surface was evaluated on

IMG. 73
Vision of the project.
By: Tredje Natur studio

the basis of current standards without affecting or limiting the existing traffic. The new green areas are conceived as green roads useful for balancing water outflows and enriching urban spaces through on-site evaluations of context elements, such as: green areas that can favor shadowing, materials used for soil cover, orientation of buildings. In the end the project also considered moving the existing parking lots to the shadow areas thus favoring exposure of trees and more efficient use of sunny space. The intersection between Bryggervangen and Landskronagade is redesigned, for example, by moving the parking areas to the shadow and leaving room to the "green stream", which follows the lines of the main streets of the neighborhood and makes it possible to channel water away from residences. Green design is extended at intersections or more significant locations to define entertainment areas. Initiatives of this kind are characterized by a broad use of green infrastructures, which are contained in the Adaptation Strategies for Climate Change in Urban Environment (ASCCUE) developed by the University of Manchester (Filpa A., Pellegrini V., 2013).

Application of this conceptual idea clearly shows that something has changed. We can note how the adopted solutions, while technically using innovative and up-to-date systems, new "green" and "blue" infrastructures, etc., are working in unison on the great landscape network for the future of citizens and with peculiar attention to their needs. This future lies in a bottom-up approach and projecting, which both focus on people and not just on infrastructures. This approach, already highlighted by the concept of Smart community, is to be witnessed in the present project, where administration and designers have managed to work smartly combining enabling technologies (Nava C., 2016), capable of responding to the need of adapting to climate changes and making our landscapes safer and more livable. All this is done by combining projects and systems of Smart communities in our landscapes and working with the Smart Grid system in progress in order to make this context part of an extensive network of nodes and catalysts connected with each other.

4.3.2.2 SOUTH-EST COSTAL PARK PHOTOVOLTAIC PERGOLA, BARCELLONA
41°24'41.6"N 2°13'41.0

Location: **Explanade – Barcelona Forum, Spain**
Design: **Martínez Lapeña + Torres Arquitectos + Estayco S.A.P.**
Construction: **Drace/Dragados/Copcisa**
Client: **Infrastructures of Lievante, City of Barcelona and Sant Adria del Bésos**
State: **Completed**
Year: **2004**
Size: **4,500 m^2**

CONTEST

The city of Barcelona is considered to be, according to some recent affirmations, one of the smartest cities in the world thanks to the advanced state of implementation of smart grids, intelligent traffic management and public lighting systems that combine with other distinctive features such as technology expertise and social cohesion. One can find an excellent example of interinstitutional cooperation and smart evolution in the innovative neighborhood 22 @ Barcelona[67]. The area is considered to be an excellent example of smart urban planning, entrepreneurial innovation and application of enabling technologies.

This innovative regeneration project has created new employments, housing and live-work spaces through five knowledge-intensive clusters: Information and Computer Technology (ICT), Media, Bio-Medical, Energy, and Design (ECPA Urban Planning and Eriksson J.).

Internationally-renowned architects were invited to create a compelling skyline of landmark buildings along the renovated monumental boulevard Avenida Diagonal, which acts as the urban spine of the district[68].

The Avenida Diagonal was reconstructed within "Cerda's masterplan" (1859) for the extension of Barcelona, which was supposed to end in a square on the seaside. But it was only a few years ago that the last stretch of the dense city on the occasion of the "Forum Barcelona 2004" finally conquered

---

[67]
22 ARROBA BCN, SAU. "What is 22 @ Barcelona?" 2006. Consulted on 27 June 2017 http://www.22barcelona.com/.

[68]
See: http://safesmart.city/en/district-22-barcelona/.

[69]
Agenda 21 for Culture is part of United Cities and Local Governments (UCLG), the global organisation of cities. It represents and defends the interests of local governments on the world stage, regardless of the size of the communities they serve. Headquartered in Barcelona, the organisation's stated mission is "to be the united voice and world advocate of democratic local self-government, promoting its values, objectives and interests, through cooperation between local governments, and within the wider international community". The Committee on culture of UCLG (also known as "Agenda 21 for culture – UCLG") focuses on the exploration of synergies between local cultural policies and other areas of sustainable development. (www.agenda21culture.net).

its access to the sea, which became possible to some extent because a part of the implementation of the "Agenda 21 for Culture"[69] takes place at the intersection between the Avenida Diagonal and the East coastline. (Ferretti M., 2017)

PROGRAM

It's just at the end of Avenida Diagonal, at this intersection, that the project of PhotoVoltaic Pergola is being placed, within the 22 @ Barcelona project. This has allowed transformation of 200 hectares of industrial area into an innovative district that offers modern spaces for a strategic concentration of intensive activities based on education, training and knowledge.

It is the most important urban transformation project in Barcelona in the recent years, and it is one of the most ambitious programs in Europe due to its peculiarities. In fact, over euro 200 million have been invested into project infrastructures through public grants.

The coexistence of innovative and dynamic companies that integrate into local ones, made up of small shops and local service providers, defines a network that has proven to be productive. The key strategic sectors are Media, Information and Communication Technologies (ICT), Medical Technologies (MedTech), Design and Energy.

Barcelona's strong capacity to develop digital and energy infrastructures favor smart city management. 22 @ Urban Lab project strengthens the idea of the city as an urban laboratory and a ground for experimenting with innovative solutions with the increasingly frequent use of editing technology (source: 22barcelona.com).

IMG. 74
Context of the city of Barcelona in relation to the project of PhotoVoltaic Pergola. By Foreign Office Architects (FOA)

4. REFERENCE

VISION

The PhotoVoltaic Pergola was part of the South-East Coastal Park in the International Forum of Cultures in 2004. The site of the park was located in a recycling area of a land-stock as Maddalena Ferretti (2017) calls it, and one of its interesting requirements was to bridge an 11-meters drop in level between the city and the waterfront beach. Due to this peculiarity, Farshid Moussavi Architecture created a series of ramps interconnected by sloped surface to circulate throughout the park. (Clusa J. et.al., 2004) (Farshid Moussavi - www.farshidmoussavi.com, 2017).

> "The resultant topography of the park presents an alternative to the traditional dichotomy between the rational geometries of French landscapes and the organic, picturesque qualities of English landscapes. It is at once complex and rational: generated by precise constraints rather than through mimicking nature."
> Farshid Moussavi - Foreign Office Architects (FOA)

DEVICES and TOOLS

The idea for the Forum Photovoltaic Power Plant was to harness renewable energy and, at the same time, to create a symbol of urban architecture for this newly developed area of the city by using photovoltaic panels as devices that would succeed in producing not only energy but also symbolic values of a place.

Some assume that there is a correlation with the fact that four years later (2008), the great esplanade that dominates this new public space was turned into the most powerful urban solar energy power plant in Spain (Clusa J. ; Marmolejo C. 2004). This marked the culmination point of an initiative to harness solar energy that began in July 2004 when the large inclined structure, popularly known as

IMG. 75
Barcellona – Esplanada Forum & Solar Panels – José Antonio Martínez Lapeña & Elías Torres Architects.
By Emili Remolins

TOOL:
PHOTOVOLTAIC PANEL

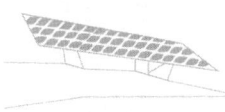

DEVICE: PHOTOVOLTAIC
CANOPY IN CULTURAL
FORUM BARCELONA 200

IMG. 76
Schematic icons about
the tools and devices of
the project.
By Giulia Garbarini

The Pergola, started to produce energy. Although it is a result of separate construction and management processes as well as independent function, it can be considered as a successful result of the use of such tools as photovoltaic panels in a landscape project and context.

Where the Pergola rises, a new type of an urban plaza is "born" at the entrance of the new marina: a dynamic, sustainable, and interactive civic canopy that plays off Barcelona's abundant natural resources. The canopy's primary element is the dimension of the photovoltaic roof, which is equipped with 2,686 photovoltaic modules with a nominal potential of 443 kilowatts. Even in the "BARCELONA SMART CITY TOUR" we can find information on how this successful energy device provided a good deal of the electrical energy and became a reference element on the Barcelona coastline and a monument to ecological commitment.

It is a south-facing, 4,500 m2 oblique surface mounted on four reinforced concrete pillars. A huge pergola receives the sun's radiation and simultaneously produces energy and shade. The total peak power of the solar panel is 1,100 kwp. It is based on silicon mono-crystalline cells.

The project's technological elements create a new set of civic ecologies that are at the same time sustainable and atmospheric. The pergola is inclined at 35° towards the south and, at its highest point, it is some 50m above sea level. When it was built, the Forum pergola became the largest urban photovoltaic power plant in Spain, and one of the five most important in Europe.

SMART PROCESS

Such tool as a photovoltaic panel in correlation with an image strategy project, ex. The Universal Forum of Cultures in 2004, not just turns into a landscaped landmark of success but is also a great opportunity for the city to take its place on the international stage and to show itself at its best. If we see it this way, it can be defined as a device that included in the smart energy category and used in both smart cities and smart landscapes, since it relates to energy potential with landscape recognition (just like landmarks), which makes it a true technology enabler. It contributes, together with urban and landscaping actions taken by the Barcelona Universal Forum 2004 project, to giving a new identity to the place.

IMG. 77
Barcellona – Esplanada Forum & Solar Panels – José Antonio Martínez Lapeña & Elías Torres Architects.
By Emili Remolins

The concept of Smart Energy is closely related to that of energy efficiency, which is based on energy savings. But it is also connected to the energy that will be supplied through the use of photovoltaic and wind power plants as well as obtained from garbage recycling.

To create a smart city and an intelligent landscape, we can work not only on energy issues in terms of engineering. Local authorities have to consider all the energy sectors that coexist in the landscape, putting as the ultimate goal the efficient use of available energy sources, in addition to the search for and effective integration of new renewable energy sources.

The Forum was to be an event of a scale akin to the Olympics. A multitude of conventions, performances, exhibitions, and even circus shows were planned with an impressive list of participants. The majority of these events, however, would not take advantage of the spaces formed through the process of preparing for the Olympics only a decade before. They would need their own new venues, requiring urban upheaval of a fresh area of Barcelona's waterfront. Vast plazas, parks, auditoriums, and convention centers were constructed along the water, just northeast of the Olympic Village.

From a smart grid perspective, one can adapt a user's behavior to power of the network, reversing today's logic; it is a matter of promoting and encouraging user's flexible behavior based on the dynamic availability of power in the network, even with the use of different energy storage systems, which allows the best use of renewable sources as these are generally not programmable.

SMART CONCEPT
The biggest difference between the efforts of urban renewal for the 1992 Olympic Games and those of the 2004 Forum of Cultures is evident in how the resulting spaces are used today. The two areas are both major public spaces of grand scale along the coast, but they offer very different experiences. The Olympic Village area has the advantage of providing services that people use everyday: bars, restaurants, clubs, and (perhaps most importantly) the beach. People can be found making use of the public spaces at any time of day (or year). The beachside promenade from Barceloneta to the Olympic Port is an attractive place to spend leisure time. The same cannot be observed if one continues north. Although the beaches continue far further up the coast (connecting to the forum area), interest steadily lessens. The large scale programs in the area have no alternate use. Herzog de Meuron's Forum Building, for example, is devastated: empty, dusty, and damaged.

The coastal area created around the forum district offers few amenities to pedestrians, even if the landscaping project designed by the group FOA (Foreign Office Architects) is a valuable and accurate landscape project. What was left after the 2004 events are sprawling empty spaces of a scale useless for much else than future large events. The result, unsurprisingly, is a wasteland of plazas. Worse yet, the unpopularity of these plazas spreads to the parks and beaches that flank them (to the point of seeing them as dangerous). The built environment here is clearly well-designed, with dynamic forms and artistically undulating hardscapes, but it lacks urban design to make it usable outside of an occasionally planned event (Rotch Traveling Scholarship, 2011).

Solution would be to work smartly with the combination of smart energy already available on the site and smart grid

IMG. 78
Barcellona – Esplanada Forum & Solar Panels – José Antonio Martínez Lapeña & Elías Torres Architects.
By Emili Remolins

architecture to make this context part of a network of catalyst nodes. Smart grids have developed drastically in the last few years, consisting of a network of information and an electrical distribution network that intelligently integrates behaviors and actions of various users connected through the exchange between Generation centers and points of use and through continuous dialogue between individuals, who can also play a dual role of consumers and producers (OICE, 2017).

4.3.2.3 VÈ-LIB E AUTO-LIB. PARIS
48° 51' 12 N - 2° 20' 55 E.

CONTEXT
Paris is the most populous city in France, and within the European continent it is ranked third in size behind Moscow and Istanbul.
This fact is of interest to us because it allows us to understand that a metropolitan area of such dimension is in close correla-

tion with public and private means of transport. Consequently, the strategies and forecasts for such a metropolis are to be taken in consideration within the European targets regarding CO^2 reductions and actions in relation to climate adaptation plans, which the city of Paris wishes and is going to adapt. The metropolitan area of Paris grew largely in the twentieth century, leading to a significant increase in the number of inhabitants in its urban area (agglomeration and periurban ring) from 10.6 to 12.4 million proportionally over the years. (Corporate Vehicle Observatory, 2017).

It seems inevitable that air pollution has always been and still is a health concern for the inhabitants of Paris. This topic has motivated the creation of the Airparif surveillance network in 1984 and, since 2001, it introduced (thanks to the reduction policies) the use of vehicles and consequently the reduction of pollutants in the urban context. The above mentioned modifications also reshaped the existing urban density, which was three times that of London with taller buildings, fewer townhouses and smaller green spaces (2,300 hectares of woodlands) (Corporate Vehicle Observatory, 2017). These modifications, even though they may seem trivial, are to be considered in the context of the last intervention of 'regaining the green in the city of Paris, which dates back to the creation of Parc de la Villette in the 1980s, while the current re-conquest of green spaces is very recent.

IMG. 79
Photo: The Parc de La Villette in 1995 following the completion of architectural and urban planning works carried out during François Mitterrand's presidency

Paris presents a fervent context of innovation and ongoing changes aimed at converting it into a Smart City, at combating climate change, but above all at transforming it into a resilient and innovative European capital. Sustainable urban mobility, districts and built environment, integrated infrastructures, energy and smart grid processes, ICT and transport, but also Focus on Citizens, integrated planning, policy, regulation and management, standards, guidelines, performance indicators and metrics, business models, supplies and financing - all this has contributed and continues to contribute to transforming the context into a smart one, rich in successful cases and potentialities (Zaza O., 2016). The main objective of the city of Paris is to respond to these challenges rationally by mobilizing collective intelligence, from citizens to industry, through public actors, by trying to involve local ones and facilitate their activities through identifying good practices, ideas, innovations and needs in order to create gradually appropriate and effective public policies.

In the field of intelligent infrastructure and mobility, the city has set up a number of networks with the prospect and aspiration to create an urban, flexible, and high-efficiency environmental model, which combines different services for the city and creates new ones. Firstly, this means working by using information in an integrated way, exploiting the synergy between all components, both in terms of energy and mobility. Convergence will be needed for data management with a dynamic, standardized and transactional information system in order to facilitate the process of sharing and processing data to ease participation, enjoyment, and cooperation of the citizens. Since 2007, the city of Paris has set an ambitious climate action plan, which demonstrates that the municipality is already interested in alternatives to individual polluting vehicles in terms of mobility and, at the same time, from the energy point of view, Paris city wants to create an energy grid system (smart grid) in order to optimize every element of the chain. This approach involves a different use and sorting system of energy and its production as well as the use of different resources in the nearby area. Therefore, this plan will ultimately lead to the approach of today's society towards a sustainable community. In order to ensure that this is done, as stated earlier, the city of Paris will work on network sessions, covering a large smart grid project "Grid4eu" funded by the European Union (http://www.grid4eu.eu/), winner of the ISGAN Award (the world's most prestigious award in the field of Smart Grids) in 2015. It is a large-scale demonstration project of advanced intelligent solutions, which sets the foundation for tomorrow's power grid by bringing together a consortium of six European energy distributors (OICE Smart City group, 2017). The project aims to put in place existing networks in different sectors such as renewable energy integration, electric vehicle development, automation, energy storage, energy efficiency and reduction of load with guarantee of scalability and replicability in Europe.

An exemplary illustration of an already built and running smart scale grid is the one present in Issy-les-Moulineaux, business district near Paris. The smart grid, developed there in 2012, meets the needs of 4,500 inhabitants, who live in 1,500 apartments on a 160,000 m² surface area[70] (OICE Smart City group, 2017).

As was mentioned earlier, within this smart grid project stands out a component related to vehicle and intelligent mobility in agreement with various experts of the sector, who stressed the need for a more sustainable urban transport system. In 2011, at the International Transport Forum, a threefold worldwide increase in the number of cars was predicted (from 850 million to 2.5 billion by 2050 (OECD / ITF, 2011). As a result, electrical mobility, modes of sharing and automation have received a major boost, here we could mention Paris Bike or Carsharing as an excellent example. They are based on a system that allows access to self-autonomous mobility without consumer being in possession of the car (ACEA, 2014). In a context such as that of Paris, giving consumers the option of renting cars on a short notice and when needed, paying only for the time they use a car and for the distance they have traveled has proved to be a remarkable success.

PROGRAM

Autolib' is a highly successful sharing service with a high frequency of use. As of 2013, most subscribers used it more than once a week, while most users who already owned a public transport card benefited from it less than twice a month. As of the end of 2013, the share of regular subscribers to Autolib' increased considerably, while the use of private means of transport decreased from 73% to 62% (Louvet N., 2014).

Innovations that change the situation in passenger transport require effective public-private collaboration (Dowling R., Kent J., 2015 & Osei-Kyei R. Chan A., 2015). On the one hand, many cities adopt a goal of promoting car-free spaces as a step towards a smart city strategy. Nevertheless, getting rid of cars requires providing citizens with access to reliable and dense transportation networks. On the other hand, private companies aim to market innovative transport products and services. However, for the most part, they fail to reach mainstream customers. How can public and private actors successfully manage changes in urban mobility?

[70]
FESR, Enel Distribution, Iberdrola, CEZ Distribuce, Vattenfall and RWE El Distribution.

IMG. 80
See the articole: Project goes live article with 250 vehicles on Parisian streets.
By Gavin Conway on December 5, 2011 5:16 PM.
Photo: AP Photo/ Christophe Ena

Changing driving habits requires creation of new organizations, processes and tools.

The Autolib' project for the city of Paris was made public by Bertrand Delano in January 2008, two months before his re-election as mayor of Paris. Implementation of an electrical charging service was added to his political agenda since the launch of Velib bikesharing program in 2007. Autolib' became a response to France's commitment to reduce carbon emissions by 20% by 2020. Autolib' benefited from following previous pilot projects, such as electric charging project, Liselec, in La Rochelle in 1999. After countless changes in technological and private-public policies, the Autolib' project was launched on December 5, 2011 with 254 cars and 256 stations distributed in 41 member cities. The synergy between Autolib' Métropole and the Bolloré Group, as well as the agility of "commanding spirit" and strong involvement of Vincent Bolloré, CEO of Bolloré Group, led to a rapid expansion of Autolib'. Société Autolib' and SMA leaders communicated through regular meetings of the SMA Monitoring Committee. Société Autolib', the delegate, produced annual reports for SMA, including financial reports and reports on quality of services. (Terrien C., Maniak R., Chen B., Shaheen S., 2016). The literature on public management of certain services illustrates that governments are able to protect innovations through public tenders, tax incentives, or subsidies (Kemp R., Schot J., Hoogma R. (1998), Smith A., Raven R. (2012)). Local governments thus can model charging services through parking regulation (Dowling R., Kent J., 2015) and, for example, lead to the success of services such as carsharing by stipulating partnership agreements between private carsharing companies and local government . The key success factor is a stable and lasting public-private relationship (more than 10 years). However, the literature on public policy leaves relatively uncovered the way local governments adapt and build long-term relationships with private actors, but in a context like the one of Pari,s we can surely find an example of success that we can make repeatable if we imagine it installable in other contexts. The Autolib' study case shows that carsharing, along with contributing to the achievement of environmental standards in cities such as Paris, can lead to creation of new organizational structures. One of the examples is Autolib' Métropole – public meta-organization – in which members of the city council, public and city administrations participated and contributed. In this way, it can be identified as a public-private interface that ensures its transparency and considers factors and issues not only from the top but also from the vision of everyday needs. Case studies suggest that carsharing projects have interrupted existing public and private organizational structures. As a result, private actors have created new organizations. City departments and municipal councils had to collaborate and reach approvals, which slowed down the decision-making process. Within Autolib', a new hybrid organization, Autolib' Métropole, was created to close the gap between public and private sectors. This result is similar to many of the new types of public-private partnerships that have emerged to launch public bikesharing services, leading to hybrid public and private business models (Shaheen, Martin, Cohen, & Finson, 2012).

Unfortunately, over the years Paris has encountered heavy traffic congestions, severe pollution, and not just in the center of the city, but also in areas outside the Île-de-France. In order to overcome this issue the city's infrastructure system had to face considerable transformation. Paris adapted expansion and adaptation programs that led to the provision of infrastructures capable of displacing 4.1 million passengers by subway every day, to reach 1,479 million of trips in Paris every year, and 48 of average trips per second. Thus, Paris began to develop various public transport or sharing systems that would allow citizens to move comfortably, quickly and with low level of environmental pollution. Among these systems are: 23000 velib (public bicycle sharing), 400 Autolib' (shared by public electric vehicles), 16 subway lines, 300 metro stations and 597 km of bus lanes (Corporate Vehicle Observatory, 2017).

On January 15, 2017, Paris became the first city with a limited traffic area and along with other types of initiatives, such as the previously illustrated smart mobility systems, it seeks to reduce air pollution by combating it through various initiatives and measures implemented by the government or by the city of Paris.

As from January 15, to circulate in a designated area when the pollution is reaching its peek one needs to have a specific Crit'Air adhesive on the windshield. The sticker shows the level of pollution produced by each vehicle and is charged 4.18€.

The most polluting unclassified cars are unable to get this adhesive and are automatically banned from the designated area during the weekdays from 8 AM to 20 AM. The adhesive is printed based on the registration number of the car.

Today, in Paris, Autolib' supplies 600 charging stations, but there is also a new Betlib companion, offering 180 public recharging stations for EV wishing to expand even more inside and outside Paris in the coming years.

VISION

The vision of this project was well illustrated by the director of the Syndicat Mixte Autolib' (2010), who described its birth, evolution and concept. Everything was born from the ambitious vision of the Mayor of Paris, who proposed a mobility solution covering a coherent territory in terms of density, variety of land use and travel needs. 41 Municipalities of the Île-de-France decided to cooperate in a public institution called "Joint Association Autolib'", and to launch consultations with private companies. They will number fifty in a few weeks.

Preliminary studies have determined that a design service to around 3,000 vehicles was good to start with. These 3,000 cars will be distributed in 1,000 to 1,200 stations with an average of 6 vehicles per parking point. The very large dispersion of the stations strengthens the network dimension and creates a close connection with users.

The aim of Autolib' and the general vision, which guides the project, is often summarized in terms of its environmental benefits, primarily in what regards the choice of electric vehicles in terms of reduced emissions of greenhouse gas. It should first be recalled that the main environmental benefits of the project are expected in terms of changing mobility patterns of users arising from the change of engine, this way, they are not limited to the environmental benefits of electric vehicles. The primary source of annoyance reported in opinion surveys in urban areas is noise, and one of the benefits of electric vehicles is that they are perfectly quiet. The second goal of Autolib' is to increase the mobility of those Parisians, who do not have access to private vehicles, usually for economic reasons. In fact, in large cities like Paris there is a very low percentage of car owners in the city center. This way, Autolib' is also addressed to all those families or individuals who, for various reasons, cannot afford a private vehicle. By changing the traditional contrast between individual movement and transport, Autolib' will bring new fluidity into the design of mobility. In this sense, it must not only be a new mode of transport performance, but also one that helps to rethink the services provided to users of other modes of transport. It is unthinkable that operators are unable to agree on a "unified ticket" for mobility in the city not connected to the chosen mobility solution.

Autolib' does not only change the prospect of public transport policies. It challenges the traditional strategy of large automobile companies, or rather it challenges the deep rooted sociological model of a car.

The challenge for the Western automotive industry and public municipalities is very simple: both must reform their strategic positioning, otherwise they risk to be overtaken by new competitors or by Asian manufacturers, mainly Chinese and Indian, who dream of gaining their advantage by occupying the almost entirely "blank page" market of electric vehicles. Without forgetting the operators of transport services, who are striving to obtain a new mobility system and simultaneously are relegating the role of the automobile industry as just one more provider and enabler of these new services.

IMG. 81
Autolib by faberNovel
2009.
Lead photo by Farber Novel

Therefore, it is not surprising to find among candidates for Autolib' the two public transport operators RATP and SNCF, grouped on one side with Avis and Vinci, and on the other with Veolia, a new entrant, Bolloré, ADA car rental company, and also a subsidiary of the G7 group, the city's biggest taxi operator.

DEVICES AND TOOLS

To date, we can state that Autolib' has become a world-renowned model of electric mobility and a reference in terms of smart cities.

Encouraged by strong demand and its growing success, Autolib' proceeded with its expansion, after the first installation, which was growing exponentially with more than 830 stations and 4,300 charging kiosks throughout the region, and 300 additional kiosks by the first half of 2014. (Paris Region Economic Development Agency, 2013). The main innovation of Autolib' in terms of smart mobility is in the energy management system, which was designed to be able to trigger a charge specifically at times when the power grid has its lowest demand.

There is also a smartphone app, which makes it possible for the whole system to get featuring outstanding functionalities. Initially developed in 15 cities including Paris, Autolib' now involves 58 municipalities. (Paris Region Economic Development Agency, 2013).

The vehicle: we hope that the vehicles that were chosen will perform well, both in terms of comfort, safety and reliability, and also when it comes to cost, both of purchase and overall operating costs (Autolib' Métropole, 2016). The network/service package: the biggest advance that Autolib' is presenting is not just the type of vehicles it uses, but above all the concept of record-setting and large scale of one-way carsharing. The phenomenon of carsharing with just a couple of recent exceptions that are still work in process (Ulm Germany and Austin Texas) has not been this successful in more than one thousand places in various parts of the world (Marty S., 2010).

The success of the extremely popular Vélib' public bike service from its inauguration in July 2007 opened up a new space for mobility between the traditional public transport (metro, regional rail, trams, buses) and private modes of travel (car, bike or walking): there is now room for "individual transport". Autolib' with its name claims to be linked directly to these new forms of urban mobility. The concept is very similar to Vélib': it would make an accessible self-service and "direct evidence" available to users of electric vehicles, that is to say by offering the option of rending the vehicle at a station different from where it was borrowed (Autolib' Métropole, 2016).

Autolib' also illustrates new potential sources of value creation in the ICT business model, beyond the benefits of the service to the retail customer. By collaborating more closely with the electric utility and DNO and ErDF, the Bolloré company running the service can improve energy demand management for its vehicles (Weiller C. 2012). Indeed, a parked set of electric vehicles that are connected to the charging network and a data network can be charged in a "smart" way, at times that make use of spare capacity in the electric grid and even that benefit it. EV batteries can offer storage services for secondary energy markets (regulation and ancillary services), particularly if they are aggregated in large battery stocks after their "life" in cars (Kohrs R. et al., 2012).

Many cities in the world (including Oslo, Barcelona, and Los Angeles) are following suit in offering electric mobility services. The rising popularity of car-sharing services and electric vehicles suggests that besides being a pioneer in the field of mobility Autolib' is also a recall of smart mobility combined of different tools that serve the community intelligently and thus become an smart enabling means of locomotion, as well as a new step towards awareness of the environment.

IMG. 82
Autolib and Velib © City of Paris

SMART PROCESS

The city of Paris received the trophy of the electro-mobile territories presented by AVERE France on December 17th, 2014. This award rewards the City's commitment in the campaign against air pollution and its action in favor of electro-mobility. The topic of mobility in Paris as well as in other cities that are (or aim to become) smart cities has common features and issues of both contemporary cities and landscapes. It also implies institutional decisions in the field of infrastructure, policy and climate adaptation plans that are not only local ones, and that have strong social and environmental implications, which make landscapes more adapted for innovation and enabling contextual devices with great potential and thanks to which they are able to fit to the required needs. Smart and sustainable mobility in relation to reductions of CO^2 emissions and to the quality of life in urban environments is mainly seen as an instrument, a device, which is made available to citizens in order not only to ensure their "right to the city" (OICE Smart City group, 2017), but also to provide its fastest, safest, and most enjoyable use. This process is conducted with a view of making our landscape smart, innovative and ready to accommodate the beliefs of the present society. This mechanism manifests itself in making it easy to connect urban areas and "approaching" suburbs to the center, urban landscapes to rural ones, and in creating a network of intangible infrastructures, the way it often happens in the logic of "smartness", thus promoting social and landscaping inclusion.

Processes implemented in the perspective of smart mobility are mainly associated with urban rather than country life (although the Wind Cycle, which is a smart example in the cycling industry, crosses different Italian regions having the same aim and project). It is becoming a key issue that links together the subjects of energy saving, sustainability, innovation and technology at different levels of scale, landscape and cities. Just as in a network that manages to express itself through physical infrastructure, which unites various layers that compose urban and landscaping systems.

The smart process that one aspires to achieve in these mobility-related processes, the experiences applied and developed to date, all that in trying to actualize the theoretical concept of "intelligent city/landscape" have brought to light various critics and issues related to mobility. These issues, however, as we have already pointed out in this paper, might function as a launch pad in overcoming various weaknesses and in promoting cutting-edge initiatives. Some of the problems that smart mobility processes face and to which they try to "respond", are, for example, increased urban traffic congestion and out-of-date vehicles; anthropogenic emissions of greenhouse gases (nowadays, in Italy, CO^2 emissions caused by transport, depending on the research, vary from 24 to 35% of the total, and our country is in the top-list of countries with the amount of cars per person); energy consumption. Indeed, we can note that mobility has an incidence of over 30% of the total energy consumption on the planet. We can therefore say that use of private cars in urban areas alone is responsible for at least 10% of all energy consumption, and that current average traffic in Italy is over 30% higher than in 1990, energy consumption and greenhouse gas emissions grew at an average annual rate of 3%[71].

Another factor to be taken into consideration is noise pollution, which can be alleviated by use of electric cars. Urban traffic severely affects the quality of life and safety of inhabitants, as reveals the first report on health impact of noise in Europe, published by the World Health Organization / Europe (2011). In addition, the last but not least important is the safety factor. The phenomenon of road accidents has significant implications, both for infrastructure and public health. In fact, it is estimated that about a quarter of nearly five million violent deaths in the world is caused by road accidents.

SMART CONCEPT

The concept behind the success of intelligent mobility, as mentioned above, is collaboration of public and private administrations, aimed at achieving more substantial global emission reductions and developing the means of public transport, which respect the environment but are also flexible to the needs of inhabitants.

In this section we have explored the emblematic case of Paris, which proved to be an example of smart mobility, especially with regard to the private mobility of low-emission cars, which can be united in a net and become nodes of distribution or

[71]. See: http://titano.sede.enea.it/Stampa/skin-2col.php?page=eneaper-dettagliofigli&id=47

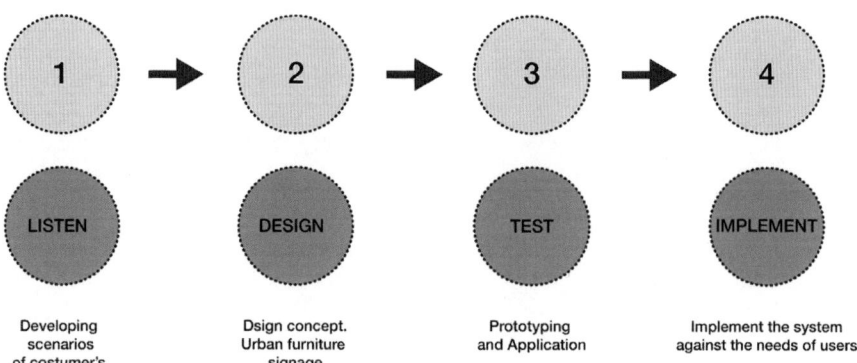

IMG. 83
Autolib diagram by Giulia Garbarini

accumulation in the network through various platforms, effectively inserted into the network of a smart grid.
However, many more examples and dynamics of smart mobility are being developed on the world level, all of them having a common concept of providing environmentally sustainable services at community's disposal. Some important examples of such practices, different from Paris, are to be found in Hong Kong's Smart/mobility cards, which can be used on trams, buses, ferries, subways, and high-speed or long-distance trains. Distribution of these cards shows significant results as 84% of the six million people in Hong Cong move by public transport, bicycles or on foot. Another example could be found in Amsterdam, where collaboration between the Municipalities, the Promoting and Supporting Agency for Research and Innovation of Economic Activities, and Gas and Electricity Operators aims to reduce the emission of CO^2 by 40% by 2025. This sort of collaboration has led to activation of bike sharing and rental systems, charging stations for electric vehicles and boat jacks, and to realization of the first commercial, participatory and energy-efficient (thanks to smart meters and smart plugs) "smart road" in Europe, and of energy saving systems for street lighting and tram stops.
This way we could state that the key concept of smart mobility is not only that of sustainability, but also and above all it is

the idea able to integrate telecommunications and computing with transport engineering, planning, design, maintenance and management of transport systems.

In today's landscape, mobility is based on the concept of social inclusion, which implies giving citizens and city users (tourists, commuters, etc.) easy access to sustainable modes of transport and alternative travel options, but also to personalized information in real time becoming an active part of the mobility offer. With the spread of mobile Internet-connected devices (smartphones, tablets, etc.), citizens become "sensors" and sources of information for the community, providing dynamic feedbacks on the status of traffic and transport (travel alternatives, road events, conditions of the means of transport etc.).

IMG. 84
An Autolib rental kiosk in Paris.

4.3.2.4 ECOGRID BORNHOLM
55° 9' 37 N - 14° 52' 0 E.

CONTEXT

The island of Bornholm is a Danish island situated in the Baltic Sea, about 160 km east of Copenhagen and about 37 km off the Swedish coast[72]. Together with the nearby island of Christiansø it is the easternmost island of Denmark. The island is part of the Danish society framework of an exemplary kind, especially in terms of tools, devices and policies that were adopted across the island to make it self-sufficient and energetically avant-garde.

Being part of Denmark's political, social and "smart" context, Bornholm has acquired its features through a long tradition of involving many different stakeholders in decision-making and planning processes regarding environmental and urban development. Among other things, this led the country to become the first state in the world to pass an environmental protection law[73]. This tradition of holistic and inclusive planning continues today and is one of the main reasons of why Denmark is considered a smart society.

It is important to know that the Danish political approach supporting green solutions is both ambitious and stable. These solutions are also expanded on the island of Bornholm. The 'State of Green' consortium, which is a public-private partnership, unites all leading actors in the fields of energy, climate, water and environment under the common plan of transforming Denmark into the first carbon-neutral country in the world by 2050. Furthermore, the "State of Green" consortium fosters relations with international stakeholders interested in learning from Danish experience. Denmark's role as a leader in promoting the green growth economy is recognized internationally, and the country has been ranked in the top two of the Global Green Economy Index in 2010, 2011 and 2012[74].

We live in a time of smart cities, which are hubs of smart devices embedded in their infrastructure as smart meters, sensors, cameras etc. Moreover, a lot of people believe that words like "technology", "design", "creativity", or "smart" characterize big cities, where serious problems should be solved

72-73-74.

Consulted August, 20. 2017 https://it.wikipedia.org/wiki/Bornholm

IMG. 85
EcoGrid logo.

by powerful companies. Bornholm proves that this approach is wrong: even if it is a small island, its administration and population have decided to show that innovation, solutions and development can take place anywhere at any time and be conducted by anyone.

Bornholm is the first place in Denmark to face the new era, and in addition to being a common incentive in the country it is also a symbol of perfect administrating. Bornholm had experienced crises and financial downturns, thus being ahead has not always been within easy reach. But in 2007 the island took a crucial step. A widely representative group of 48 people from the Bornholm community boarded one of its biggest ferries to engender new hope and direction and develop a new strategy for the island. As the ferry was plying the waters surrounding the island for 48 hours, the group developed a vision which we now call "Bright Green Island".

The urban country life on Bornholm attracts thinkers, doers and makers and offers them a green creative playground. The municipality had a wish to turn a previously fringe area of Denmark into an asset, an island replete with green initiatives, projects and enterprises.

Bornholm has chosen a bold new direction and will keep on turning all obstacles into opportunities, developing new solutions to meet major challenges.

PROGRAM

The EU electricity and energy efficiency directives clearly state that the demand side should have access to the power system on an equal basis as that of generation. Also, the European Commission requires that the member states develop clear rules (Network Codes for smart grid development) for system operation and market functions that foster demand-side participation in power market balancing.

From the very beginning, the ambitious objective of EcoGrid EU was to develop and demonstrate on a large scale a generally applicable real-time market concept for smart electricity distribution networks with high percentage of renewable energy sources and active user participation. It should thereby reduce the need for costly flexibility on the production side and/or compensate for traditional balancing power and services from conventional generation displaced by generation based on renewable energy sources.

The very fundamental program of EcoGrid EU is to balance power systems by repeatedly issuing a real-time price signal for flexible resources to respond to. The price signal will be continuously updated in order to keep the power system balanced by increasing the price when there is a power deficit in the system, and vice versa.

To test the EcoGrid EU concept, smart home equipment was installed in a large number of households. Customers were able to either manually respond to real-time prices, or received equipment that controlled their heating system to respond automatically to price signals.

The EcoGrid in Bornholm turns out to be such a test area of this program, where a micro smart grid is applied to the island's context and in this case the developers are able to test the program in a controlled manner. The EcoGrid in Bornholm has proven that the customers reacted in a way that helped balancing the power system following a real-time price signal.

IMG. 86
René Kamphuis, TNO, the Netherlands at the IEA DSM workshop in Utrecht, the Netherlands on April, 24, 2013.

Project database: Ecogrid Bornholm

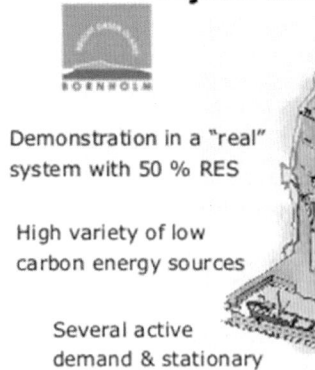

Strong political commitment & public support

Demonstration in a "real" system with 50 % RES

Operated by the local municipal owned DSO, Østkraft

High variety of low carbon energy sources

Several active demand & stationary storage options

Eligible RD&D infrastructure & full scale test laboratory

Interconnected with the Nordic power Market

Also, there is a significant peak load reduction potential: activation of flexible consumption with a five-minute real-time signal reduced the total peak load of EcoGrid EU participants by approx. 670 kW (Grande O.S., 2015).

In a perspective of replication, customer involvement is the key to success. Moreover, an important task was keeping participants involved throughout the project. And another precondition for wider smart grid expansion in general and the EcoGrid EU real-time market in particular is the design of immediately available equipment that is specifically created for automatically providing power system services to the TSO or DSO upon receiving an external control signal of any kind (market or technical) (Gantenbein D. et. al., 2012).

VISION

The pilot implementation and demonstration on Bornholm are only just starting, thus more time is needed to report on results.

However, the prospects for EcoGrid EU to create a "win-win" situation, enabling small and large electricity customers to save money on their electricity bill, while also relieving the power system, are good. And in the longer term, this will also reduce investments in grid reinforcements and new grids. The savings at a European level have not yet been estimated, but the Danish electricity sector and En- erginet.dk have calculated a direct socio-economic saving of at least DKK 1.6 billion when using smart-grid solutions in Denmark. Furthermore, an extra bonus will be the environmental benefit Denmark will achieve by improving the integration of environmentally-friendly electricity and power savings.

Gantenbein D. et. al. (2012)

The prospective for Bornholm to be turned into a Green Island has four cornerstones: Sustainable Business, Good Living, Smart Island and Green Destination. These four categories are organized in a way that they can be viewed as one image with four autonomous sectors. By complementing one another, they comprise a combined vision.

The municipalities of Bornholm and the commission of EcoGrid project have accumulated vast experience and know-how in each of the four categories and they wish to share it with other islands and regions around the world, which are aiming at green and sustainable development.

For a common future of the whole Europe Bornholm Island hopes to become a 100% sustainable and carbon-free community by 2025. A community, which creates local, sustainable and eco-friendly solutions, growth and new businesses. A community which shows the world how a small island can take qualified steps towards coping with urgent global challenges such as scarce resources and greenhouse effect.

The Danish island of Bornholm is becoming one of the most popular test sites in the world for proving new green technologies. Because of its geographical location and well-developed water, heating, and electricity systems, the island is ideal for testing electric cars, solar panels, smart buildings, and intelligent systems, all of which will play an important role in future Smart Cities. Recently, the Bornholm Growth Forum formed a partnership with Vaeksthus Greater Copenhagen in order to develop a unique calculation model (Bright Green Island, 2013). The model will make it possible to transfer the test results from Bornholm to other similar communities around the world. As an example of unique opportunities on Bornholm, Østkraft, the energy utility company on the island, is involved in several international test projects aimed at discovering how stability of supply can be maintained once a far greater percentage of power production comes from renewable sources.

One example is the €26.8 million project, EcoGrid EU, which has helped to install computers in 2,000 households on the island, testing smart management of electricity consumption. The project actively involves end-users in the electricity market so that it becomes possible to control consumption based on the price of electricity. A specific household computer makes it feasible for the household to switch selected devices on and off based on the price[75].

75.
Consulted Agust, 20, 2017 www.bornholm.dk

IMG. 87
Green Island strategy
by EcoGrid Bornholm

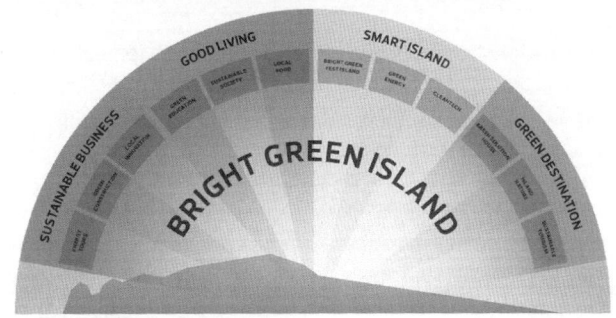

BORNHOLM

DEVICES & TOOLS

Inserted in a broader perspective, smart grid and European management system, the island applies in its context a range of Tools and Devices that aim at different objectives, which are well-synthesized and explained in the document "Reference for the study of the overall project" ("Bornholm Bright Green Island")

The main areas where we can encompass three devices are: SUSTAINABLE BUSINESS, GOOD LIVING, SMART ISLAND, and GREEN DESTINATION.

Below we will go back to a univocal set of schemas and give a brief description of each of these macro groups.

The first category that was developed on the island is SUSTAINABLE BUSINESS, which has several different subcategories within it: energy tourism, green construction, and local innovation. As mentioned in the above quoted document, this macro group is the way to show that the future belongs to those who invent it, and that Bornholm has undergone several business crises and financial downturns over the years, but each time has managed to recover. Since the Bright Green Island strategy came into life in 2007, sustainability, innovation and green technologies have become an increasingly widespread common denominator for Bornholm's business community.

Energy Tours have attracted energy tourists to the island from all around the world, including countries like Thailand, Germany, South Korea, Japan, Poland, Sweden, Hong Kong, Vietnam, China, the USA and Russia. There are a total of fifteen different Energy Tours which can be combined on demand.

Green Construction is expected to create 100 jobs by means of a focused, large-scale effort to create energy-efficient renovation. Close to 10,000 households will be encouraged to assess their needs for energy advice and guidance in collaboration with the island's utility companies. The aim is to provide 1,000 energy advice sessions, as this is the first step towards energy-efficient renovation. In terms of social sustainability, the Bright Green Island vision has proven its ability to inspire the development of new business models which provide meaningful employment to people with disabilities or other challenges that exclude them from ordinary employment.

The second category is GOOD LIVING which has several different subcategories within it: green education, sustainable society, local food. This macro group regards the green conversion process and goes beyond renewable energy and technological solutions. It is crucial for each individual to adopt this trend. It is important to teach ourselves and our children to think sustainably, to change our habits and to motivate businesses and industries to produce greener foods. The Project Bright Green Campus influences the study programs offered by Campus Bornholm in a green, sustainable and innovative direction. The municipality believes that people who care for their local environment also want to take good care of it. And we believe that involvement and possibility of influencing local decisions is the right way to support the vision of a sustainable island community.

The third category is SMART ISLAND, which has several different subcategories within it: Bright Green test Island, green energy, cleantech. This macro group hosts a number of R&D projects involving electric cars, solar cells, energy efficient construction projects, and development of an intelligent electricity system, each of which aims at reducing carbon emissions. The interest in testing and trying out new concepts and products on Bornholm is primarily prompted by two factors: Bornholm's power grid is connected to the rest of the world by only one network: a long submarine cable that goes Sweden. This fact makes it possible to isolate the island for electricity purposes and monitor the volume of electricity imported and exported. Thus, the impact of changes and new factors can be determined with unprecedented accuracy.

The fourth category is GREEN DESTINATION which has several different subcategories within it: Green Solution house, island's nature, sustainable tourism. Within this macro group Bornholm's biggest attraction is its widely varied nature. The forces of nature cover Bornholm with snow, create turbulent waves, cause heavy storms and provide moments of solitude. But it is these very same conditions and surroundings that attract people to settle here, bring new visitors and fascinate tourists who keep coming back year after year. The serenity of Bornholm's countryside provides ideal conditions for relaxing, meditating and exercising outdoors. Bright Green Island is based on a desire to promote nature conservation to give people better natural living conditions. Our local efforts depict our global desires.

SMART PROCESS

Denmark is a highly digitalized society, which makes it ideal when it comes to implementing or testing new and smart solutions.
Since 2001, the Danish government together with the municipalities and the regions has been engaged in a strategic partnership to transform Denmark into a smart and digitalized society. The work revolves around a national Digitalization Strategy, which aims to create a smarter, more efficient and cohesive public sector[76]. The strategy is based on two approaches: first of all, a targeted effort is being made to digitalize communication between public authorities, citizens and businesses. Secondly, the Digitalization Strategy focuses on ways of making various Danish public authorities more digitally integrated.
Denmark's well-developed digital infrastructure supports the country's smart transition as it makes it easier for companies, public authorities, researchers, and citizens to connect and share valuable data. As Søren Smidt-Jensen from the Danish Architecture Centre says: *"We need to think more about the need for integrating sensors and different components in everything from buildings to our means of transport"* (Copenhagen Cleantech Cluster, 2012). The effective digital infrastructure combined with the need to implement smart technologies means that there is great business potential for foreign companies to test their smart products or services in Danish cities.

[76]. Enenkel, e ektivog sammenhængende o entlig sektor. The Government, Danish Regions & Local Government Denmark 2011

In order to create smarter technologies, cities and eventually landscape, Denmark is working systematically to foster new and innovative ways of breaking down the great amount of knowledge and facilitating collaboration across the society. One example of how this is done is the "innovation platform" approach, which has been adopted by the City of Copenhagen in collaboration with Copenhagen Cleantech Cluster. The aim of the approach is to explore how public procurement can be used as a driver for innovation.

Copenhagen was ranked among the top Smart Cities in the EU and we can understand that the Danish cities provide an ideal test market for new smart technologies and solutions and that the country does not only occupy the leading position within several key green technologies needed for the Smart City. Moreover, its long tradition of involving different stakeholders in the planning and decision-making process makes it an ideal living lab for smart solutions. This is supported by the fact that Denmark is a highly digitalized society, where all stakeholders are able to connect, share and collaborate in new and innovative ways. Furthermore, municipalities, companies and citizens provide a wide variety of open data, which can be used in the development of new smart technologies. Finally, Denmark is an innovative country, where new forms of collaboration across society are being constantly developed and tested.

SMART CONCEPT

Commonly and historically, most energy distribution systems are centralized around a few big power plants, but nowadays the situation is changing, or rather it needs to be changed and adapted to new technologies and new global expansions. According to the necessity of shift that is being made even without our will, we can try to adapt to the changes, one of which concerns huge and inefficient electrical grids that lose power while transporting energy from power plants to cities. Furthermore, as the supply of electricity needs to respond the demand at all times, power plants have to generate an overcapacity of electricity in order to cope with unexpected rise in energy demand. This, of course, is both highly inefficient and damaging for the environment. Secondly, the integration of new and decentralized energy sources into an energy grid presents challenges such as the way in which this energy grid is controlled. As Tyge Kjaer, Associate Professor at the Department of Environmental, Social and Spatial Change at Roskilde University, points out: "Our use of energy is three times bigger than what is realistically possible in the long run, which means that we need to develop far more intelligent energy systems. A prerequisite for this is that we take an energy system approach. We should ask ourselves how the individual source of energy fits into the bigger system" (Tyge K., Rikke L. 2015). In other words, the energy grid of the future needs to be far more flexible and intelligent. The term 'Smart Grid' refers to a new kind of energy network, which makes use of software and hardware tools to monitor and manage the transportation of electricity from all generation sources connected to the network. This in turn provides for a more flexible and less wasteful energy transmission process, which is able to integrate decentralized and local power generation sources such as solar panels, wind turbines, and heat pumps.

The European Union's concept that provides us with a program like the one of "EcoGrid Europe", which is a precursor for the island of Bornholm, aims at a fast evolution towards intelligent broadcasting networks. The fundamental goal and basic concept is to contribute to the European 20-20-20 targets.

It is trying to use Smart Communications and Smart Market solutions as key contributors to the application of Smart Grid. The overall project of EcoGrid EU is a large-scale demonstration of a complete feeding system that includes the following elements:

- total distributed grid with all the resources up to 60 kV; 28,000 customers; peak load 55 MW; 268 GWh electricity consumption; 500 GWh heat demand;
- distributed RES, including wind power (30 MW), photovoltaic (2 MW), biomass (16 MW), biogas (2 MW) and electric vehicles;
- TIC Systems, a new information architecture that allows all DER units and DR consumers to participate in the energy market;
- Smart House Appliances, Smart Counters, E-Mobility Using Electric Vehicles as an integral part of a wider system.

With the application of these devices and new technological components landscape can be observed under one single lens, one test. The island of Bornholm is a demonstration of the fact that a smart grid is able to implement common large-scale objectives (Europe) leading to global results and can really change the current situation.

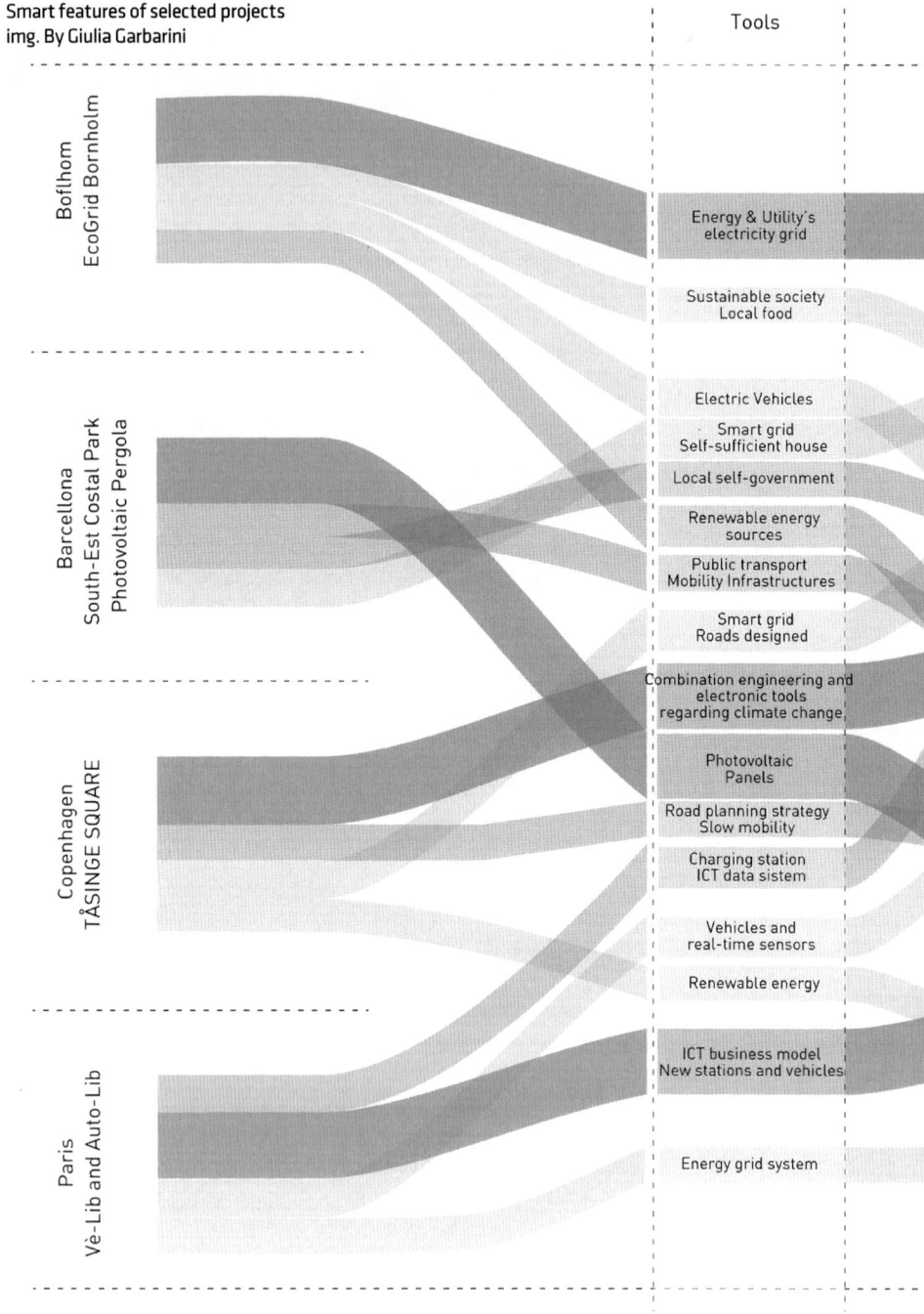

IMG. 88
Smart features of selected projects
img. By Giulia Garbarini

SMART LANDSCAPE

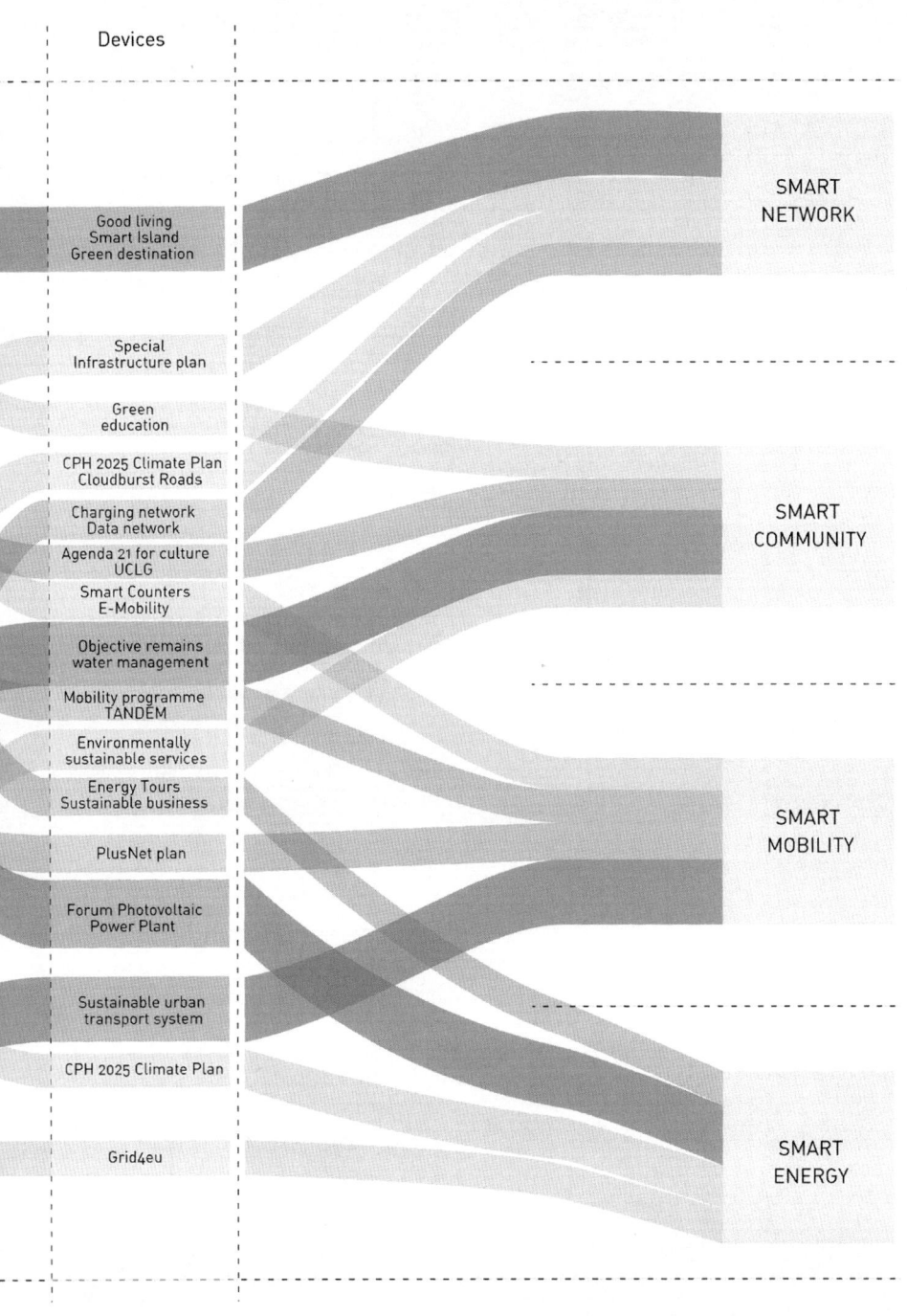

CHAPTER FIVE

5

TRANSFER KNOWLEDGE

"Smart Landscape" aims at defining new design paradigms[77] related to energy issues. It highlights the way a rigid system such as smart grid can be used dynamically if integrated and implemented with new perspectives of design, when it is not imposed from above but answers to the needs of the examined places in a targeted way. As underlined previously, today we are immersed in an inter-connected landscape, and it is from this network that "Smart Landscape" is able to take root. In this context, we can find a common place where to try to respond to environmental needs that occur especially in anthropized environments. The present book proposes to work on landscape in an intelligent way through overlapping these two "surfaces" (landscape and smart grid technology) in such a way that puts in place some tools and processes and allows a combination of these two aspects.

Today, we are observing a technologically advanced landscape (and it will become even more so in future), which uses new technology to its advantage. Nonetheless, "Smart Landscape", as already pointed out above, has to be considered smart because by leveraging technological, political and social advances occurring in different European countries it aspires to becoming a "responsive" landscape, where actors who inhabit and build places of living, manufacturing, agriculture etc. understand that there is a need to work and construct different landscapes and to use the latest technologies to make landscape system "sentient". In other words, able to understand the limits of sustainability and to intervene on landscape overcoming these limits.

Combination of these two disciplines (landscape and smartness) could get its "constructive foundations" (basis) in the architecture of micro and macro smart grids, which allow various systems to be constructed of different energy, innovative and avant-garde processes, due to interoperability of these systems, processes and devices, such as those illustrated in chapter 4. First of all, scenarios of change should be created in the examined places, not antithetical to each other, with the desire to define and build a super smart grid extended between different contexts, which are not only designed to create energy corridors, but also to promote the exchange of services linked, for example, to mobility, landscape and society. Secondly, a shared vision should be established

77. Ricci M., Nuovi paradigmi, List, Trento, 2012.

through scenarios that lead to reconfiguration of urban or landscape contexts as living organisms endowed with a capacity to define connections between different parts of the city and landscape systems.

In this sense, the objective of this book and research is not to trivialize the strength of landscape or follow a trend. On the contrary, its purpose is to assert that for some time landscape has been regarded an area to which we refer due to its ecological sense, which makes it a unitary horizon, within which different contexts, cities included, can be placed.

We want to propose an articulation for knots within landscape to promote a regenerative catalysis both of the places in question and of the surrounding areas, where by exploiting the large scale of landscape we can turn our gaze and work with different surfaces, devices and policies, thus creating a sort of realignment in which landscape replaces urban planning and architecture (Corner J., 2006). Due to the mixture of a wide range of disciplines, landscape becomes a target through which we can represent widespread contemporary cities but also build them in relation to the contexts that are interposed between one and the other.

These objectives can be summarized in a process divided into three main parts, which are based on the technology of Energy Smart Grid, and which aim at being repeatable and applicable in other contexts.

The experimental profile of this process is outlined by combination of theoretical elaboration of a Smart Landscape model (chapter 3) and application process already existing in a contexts or being elaborated. It is able, or is trying to combine concepts of landscape and smartness, as we have seen in the contexts examined in chapter 4, where there are conditions for a possible "Smart Landscape". In the places, where smart devices or "responsive" landscape projects have already been set up, it is possible to launch an application process of "Smart Landscape". The three phases of it are: operational learning (with a reconstruction of cognitive frameworks through decomposition of the context in ecosystems), interpretative learning (in which the existent circles are identified as catalysts of change, following the activation of devices that can favor the transformation), and exploratory learning (in which scenarios are seen as predictions of change

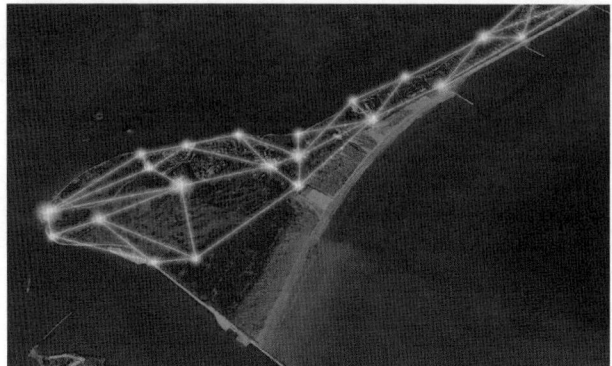

IMG. 89
New interchange processes between different systems and devices.
Sud-West Lido Island.
img. Giulia Garbarini

spatial prefiguration of which is entrusted to visions); all this is described in the book "L.I.D.O. - LEARNING ISLAND DESIGN OPPORTUNITIES ".

This combination allows anchoring and stabilization of the concept of Smart Landscape to the relative urban or rural context but simultaneously becomes dissemination of the latter. Similarly to the function of roots in plants, the concepts of smartness and landscape in relation to the reference projects and the case study have the function of anchoring, of absorbing "the nourishment" from the context and of communicating between similar "organisms".

The objective of the hypothesized process is to transform mere energy nodes into catalyzing nodes of social and landscape value, to use systems for production of renewable energy, to improve efficiency of the places in which they are installed and to collaborate with the catalysts, aiming at respecting the European resilience policies.

There are three peculiar characteristics of the research that can be transferred into different contexts in order to develop a smart landscape.

STRATEGY - Landscaping

In the current European geopolitical framework, in response to the needs made obvious during the summits on climate change and energy saving, in anticipation of the European objectives to be achieved through the possibility of joining European calls issued by the Horizon 2020 program, local governments are trying to accelerate the phenomena of adaptation and integration of energy and urban systems. They are applying the concept of anticipating climate change effects by prioritizing actions that primarily aim at adequate prevention/reduction of contemporary threats, and that may be able to exploit the opportunities resulting from these changes.

Growing awareness of climate change, along with the inevitable impacts and consequences suggests that the spatial appearance and organization of urban and rural landscapes will undergo far-reaching changes and influences.

One of the possible strategies is to alter anthropogenic contribution of this change through a multi-scale approach that seeks to create a network of knowledge and proposals. As illustrated in the "strategy of surface" by James Corner, research is channeled towards a strategy and an adaptive surface. After reading the context of ecosystems it is able to activate the existing resources by adopting foundational concepts of the micro and macro smart grid technology, which becomes a link between various surfaces that create landscape and guide us to a new concept to observe and possibly intervene on.

STRUCTURE - Energy Model

The structure of smart landscape is mainly conceived and based on the architecture of energy smart grids. It can also be divided into an ecosystem component as a key to understanding the potential and deficiencies of the context consisting of energy, landscape and society ecosystems, and an innovative component, consisting of micro and marco smart grids, from which it adopts the structure made of production nodes and exchange of values and energy, of virtual and physical connections.

The prototype proposed as part of these two components should be able to adapt, to the context through a series of actions, devices, programs or planned interventions capable of producing new levels of innovation, always referring to the European objectives.

PROCESS - Deduction

It is also possible to refer to the technology of smart grids especially with regard to management of flows. While in a traditional system it refers to management of energy flows in a flexible way ensuring production and widespread balancing of

the whole network, in the process described in this research it refers to balancing of services and resources aimed at defining various scenarios.
The objective of this process is to create prefigured visions of the scenarios that become guides of change in the examined context by activating nodes identified with the help of specific devices, thus bringing benefits to the entire system and simultaneously to other expected scenarios, which are not (nor should they be) in competition with each other, but rather form a combination of different actions or devices in different scenarios and lead to improvement of deficiencies of an ecosystem with the help of enabling technologies.

--

Finally, the question arises:
what strategies can be adapted to deal with changes caused by forces such as accessibility, urbanization, globalization, natural disasters, natural, cultural, social and industrial developments, to keep landscapes?
The answer could be in adopting and applying the concept of "Smart Landscape". We can state that prospects for future work could be aimed at experimenting and validating it (the concept) by finding a way to transform it into a real project. In fact, it is important to specify that the present book is the result of PhD research entitled "Smart Landscape: The architecture of the "micro smart grid" as a resilience strategy for landscape" elaborated at the University of Trento, Italy. Within this research the case study (situated on the Lido Island of Venice under the acronym L.I.D.O. - LEARNING ISLAND DESIGN OPPORTUNITIES) was presented as a prototype of a project. It is not a univocal model that can be applied in other contexts but a way in which one could work by relating enabling technologies and projects with landscape discipline, which leaves the field open for other theoretical experiments and diversions.
By combining theoretical and practical exploration launched on the L.I.D.O. we wanted to propose a possible reasoning and a preliminary experimentation able to use technological innovations in order to create "Smart Landscape".

Working and constructing "Smart Landscape" could mean bringing different instances and disciplines together through conceptually conflicting processes, identifying in it the possibility to catalyze different subjects, interests and resources. In fact, the word "smart", although being among the most popular today and often used only as an adjective that expresses multiple meanings, if related to landscape, it can underline the abilities and dynamics that this discipline possesses, which lets it react in a dynamic way by learning from the "context".

Being "smart" does not only mean being intelligent but also ready, witty, bright, reactive and above all it means being able to show adaptation and quick learning skills. Today's landscape should navigate within this broad spectrum of meanings adapting itself to the needs of a place, exploiting risks as potentials and responding to social and environmental needs by connecting different existing systems and disciplines in order to transform places of everyday life into CATALYSTS (a term used in its positive sense) (Oswalt P 2013), which, following the activation or the realization of some DEVICES (technological, managerial, social, urban, etc.), can favor or accelerate transformation.

BIBLIOGRAPHY

Not all the books listed in the bibliography are directly quoted in the thesis text. Some of them are books which helped me to set up the theoretical background where to start my research.

- AA.VV. (2007), "Report of the Intergovernmental Panel on Climate Change" Cambridge University Press, Cambridge, United Kingdom and New York, NY, USA.
- AA.VV. (2011), "White Paper on Smart Grids", Brussels, DIGITALEUROPE.
- AA. VV. Edit by : Ricci M., Scaglione G., Favargiotti S., Rizzi C., Staniscia S. (2015), "Monograph.RESEARCH. R.E.D.S.2Alps Designing a Sustainable Future" Trento: LISt Lab, p. 274-279.
- AA. VV. (Eleonora Riva Sanseverino, Raffaella Riva Sanseverino, Valentina Vaccaro) (2015), "Atlante delle smart city. Comunità intelligenti europee ed asiatiche: Comunità intelligenti europee ed asiatiche" Edit by Franco Angeli, Milano.
- Abdel-Aziz A. A. , Abdel-Salam H. , El-Sayad Z. (2016), "The role of ICTs in creating the new social public place of the digital era". In : Alexandria Engineering Journal. Volume 55, Issue 1, March 2016, Pages 487-493
- Agodi M. C. (1996), "Qualità e quantità: un falso dilemma e tanti equivoci", in Cipolla C., De Lillo A. (a cura di), Il sociologo e le sirene. La sfida dei metodi qualitativi, Franco Angeli, Milano, pp. 136-166.
- Anderson E., (2013). "Placemaking: A Tool for Rural and Urban Communities". <http://www.extension.org/pages/67018/placemaking:-a-tool-for-rural-and-urban-communities#.VPr5PeF3JyE> (accessed 11.09.17).
- Antrop, M. (2005), "Why landscapes of the past are important for the future", Landscape and Urban Planning 70 pp. 21–34, Elsevier, Amstredam.
- Audretsch D. B. (1995), "Innovation and Industry Evolution", MIT Press, Cambridge, Massachusetts, US.
- Augé M. (2009), "Nonluoghi, introduzione a una antropologia della surmodernita", Elèuthera, Milano.
- Assunto R. (1973), "Il paesaggio e l'estetica." Natura e Storia, Napoli, Giannini, p.176.
- Bach R.C. (2008), "Surface Strategies" som landskabsurban metode". Nordic Journal of Architectural Research Volume 20, No 2, 2008, 11 pages Bach R., Clemmensen T. (2005), "Landskabstilgangen". Nordisk Arkitekturforskning #2. http://aarch.dk/institutter/institut_for_by_og_lands- kab/arbejdspapirer/
- Barry R.G., Chorley R.J. (1992), "Atmosphere, Weather and Climate", (VI edition), Routledge, London and New York. PP.35
- Bertini I., Marrocco M. (edit by) (2017), "Distributed Generation Manager of the ENEA Research Center." explains the intelligent networks, ENEA magazine.
- Bonesio L. (1997), "Geofilosofia del paesaggio", Mimesis, Milano.
- Bonesio L. (2007), "Paesaggio, identità e comunità tra locale e globale", Diabasis, Parma.

- Bonomi A. Masiero R. (2014), "Dalla smart City alla Smart Land", Edit by Marsilio s.p.a., Venezia
- Branzi A. (2006), "MODERNITA' DEBOLE E DIFFUSA Il mondo del progetto all'inizio del XXI secolo", Skira, Ginevra-Milano
- Branzi A. (2006), "No-stop city: Archizoom Associati". Editions HYX, Orléans (FR).
- Brundtland, G.H. (1987), "Our Common Future: Presentation of the report of the World Commission on Environment and Development." In: UNEP's 14th Governing Council Session. Oxford. Oxford University Press
- Bryman A., Burgess B. (1994), "Analyzing Qualitative Data", Routledge; 1 edition
- Buccaro A. (2011), "Leonardo da Vinci. Il Codice Corazza nella Biblioteca Nazionale di Napoli." Poggio a Caiano/Napoli: CB Edizioni/Ediz. Scientifiche Italiane.
- Buonanno D. (2014), "RURALURBANISM Paesaggi produttivi", Università degli Studi di Napoli Federico II Facoltà di Architettura. Dottorato di Ricerca in Progettazione Urbana e Urbanistica_XXVI ciclo.
- Calvino I. (1993), "Le città invisibili", Edit by Mondadori, pp. 111-112, Milano.
- Camarsa G., Toland J., Eldridge J., Nottingham S., Hudson T., Jones W., Heppner K., Silva J., Thévignot C. (2015), "LIFE and Climate change adaptation" . Luxembourg: Publications O ce of the European Union.
- Caragliu A., Del Bo C., Nijkamp P. (2009), "Smart Cities in Europe", Serie Research Memoranda 0048, VU University Amsterdam, Faculty of Economics, Business Administration and Econometrics.
- Carta M. (2014), "Reimagining Urbanism", List, Trento.
- Carus G. C. (1985), "Nove lettere sulla pittura di paesaggio, in appendice a A. Sibrilli, Paesaggi dal nord. L'idea del paesaggio nella pittura tedesca del primo Ottocento", Roma, Officina Edizioni, p. 186.
- Clark C., Kent E., Kent F., Lester R., Madden K., Meynell I., Pan S., Raphael C., (2008). "Neighborhood Placemaking in Chicago" http://www.placemakingchicago.com/cmsfiles/placemaking_guide.pdf (accessed 11.09.17)
- Cook D. (2005), "Smart environments. Technology, protocols and applications", Hoboken, N.J, Wiley.
- Corner J. (2006), "Terra fluxus. In The Landscape Urbanism Reader", Edit by Charles Waldheim, 21–33. New York: Princeton Architectural Press.
- Council of Europe (2000), "European Landscape Convention." Firenze. In: European Treaty Series 176/2000 Strasburg, Council of Europe.
- Corner J., MacLean A.S. (2000), "Taking Measures Across the American Landscape". New Haven: Yale University Press.
- Corner J. (2003), "Landscape Urbanism", in Mostafavi M., Najle C., Landscape Urbanism a manual for the machinic landscape, AA press p.63, London, UK.
- Corner J. (2006), "Terra fluxus", in Waldheim C., The landscape urbanism reader, Architectural Press, p.29
- D'Angelo P. (2009), "Introduzione", in D'Angelo P. (a cura di), Estetica e paesaggio, Il Mulino, Bologna, pp.7-38.
- Daniels T. L., Lapping M. (2005), "Land Preservation: An Essential Ingredient in Smart Growth". Departmental Papers (City and Regional Planning), SAGE Publications, Pennsylvania USA. http://repository.upenn.edu/cplan_papers/25

- Gambino R (1997) Conservare, innovare. Paesaggio, ambiente, territorio. UTET Libreria, Torino.
- Deleuze G. (2007). "Che cos'è un dispositivo?". Edit Cronopio. Napoli.
- Desiato F., Giordano F., Perconti W. (2007), "I cambiamenti climatici influenzano natura ed economia". Conferenza Nazionale sui Cambiamenti Climatici, Roma, 12-13 Settembre 2007.
- Drucker P. (1954), "The practice of management." New York, NY: HarperCollins.
- ENEA, Ente per le Nuove Tecnologie, l'Energia e l'Ambiente (2007). "Rapporto Energia e Ambiente 2006". Edito da ENEA – Unita Comunicazione, Roma, aprile 2007.
- European Commission (2007), "Green Paper on Adapting to climate change in Europe – options for EU action." In: COM (2007) 354 final. Strasburg, Council of Europe.
- Fabietti. U., Malighetti. R., Matera V. (2000), "Dal tribale al globale", Mondatori, Milano.
- Faggian P. et. alt. (2008), "Climate change on Mediterranean Region", Convegno: European Geosciences Union: General Assembly, Vienna, 13-18 Aprile 2008.
- Farinelli F. (2003), "Geografia. Un'introduzione ai modelli del mondo." Torino, Einaudi, p. 232
- Farinelli F. (2015), "La capriola del paesaggio, in Quindici anni dopo la Convenzione Europea del Paesaggio 2000-2015." In: Sentieri urbani 17/2015. Trento, Bi Quattro Editrice, p.18-22
- Farinelli F. (2015), "Il ritorno del paesaggio." In Corriere della Sera ed. (December 20, 2015). Milan, Rcs Quotidiani S.p.A., p.48
- Gambino R. (2003), "Progetto e conservazione del paesaggio." Ri-Vista, Ricerche per la progettazione del paesaggio 1(0), Firenze University Press.
- Gausa M. (2017). "Land-Link. Il paesaggio come infra/intra/eco (e info) struttura territoriale." In Monograph.Research REDS 03. Flowing Knowledge, Trento: LISt Lab, pp. 324-333
- George, C. S. (1972), "The history of management thought. Englewood Cliffs", NJ: Prentice-Hall, Inc.
- Giordano V., Meletiou A., Covrig C. F., Mengolini A., Ardelean M., Fulli G., Jiménez M. S., Filiou C. (2013). "Smart Grid projects in Europe:Lessons learned and current developments".Publications office of the European Union, Netherland.
- Giordano V., Meletiou A., Covrig C. F., Mengolini A., Ardelean M., Fulli G., Vasiljevska J., Jiménez M. S., Filiou C. (2014). "Smart Grid Projects Outlook 2014". Publications office of the European Union, Netherland.
- Goethe J. W. (1991), "Viaggio in Itali." BUR Biblioteca Univ. Rizzoli, Milano.
- Gombrich E. (1960), "Art and Illusion." Phaidon, London.
- Graham, S. (2000), "Introduction: cities and infrastructure networks", in International Journal of Urban and Regional Research, Vol. 24. New York City, USA.
- Graham, S. and Marvin, S. (1999), "Planning cyber-cities? Integrating telecommunications into urban planning", in Town Planning Review, Vol. 70. Liverpool, UK.
- Gravano V. (2012), "Paesaggi attivi. Saggio contro la contemplazione." L'arte contemporanea e il paesaggio metropolitano, Mimesis, Collana: Kosmos, Milano-Udine

- Groat L.N., Wang D. (2013), "Architectural Research Methods", John Wiley & Sons Inc
- Habermas J. (1971), "Storia e critica dell'opinione pubblica." Bari, Laterza, specie pp. 119, 125.
- Hall R. E. (2000), "The vision of a smart city", presented at the Second International Life Extention Technology Workshop, Paris, France.
- Hersey, P. H., & Blanchard, K. (1988), "Management of organizational behavior". Englewood Cliffs, NJ: Prentice-Hall, Inc.
- IEA, International Energy Agency (2006). "World Energy Outlook 2006". OECD/IEA.
- IPCC, International Panel on Climate Change (2007), "Climate Change 2007: Climate Change Impacts, Adaptation and Vulnerability - Summary for Policymakers". Working Group, Contribution to the Intergovernmental Panel on Climate Change Fourth Assessment Report, April 2007.
- Jackson J.B. (1984), "Discovering the Vernacular Landscape." Yale University Press, New Haven
- Jakob M. (2009), "Il paesaggio", il Mulino, Bologna.
- Koolhaas R. & Mau B. (1995), "SMLXL". Monacelli Press, NYC.
- Koolhaas R. (2006), "Junkspace. Per un ripensamento radicale dello spazio urbano", Quodlibet, Macerata.
- Küster H. (2010), "Piccola storia del paesaggio. Uomo, mondo, rappresentazione.", Donzellini Editore, Roma.
- Lynch K. (1981), "Il senso del territorio." tr. it. "A cura di Maria Parodi,Il Saggiatore, Milano.
- Lyster C. (2006), "Landscapes of exchange: re-articulating site", in Waldheim C., The landscape urbanism reader, Architectural Press p.221, New York, USA.
- Mandelbrot B. (1987) "Gli oggetti frattali." Torino, Einaudi.
- Mau B.(2004), "Massive Change." Phaidon Press, London, UK.
- Miles M., Huberman A. (1984), "Qualitative Data Analysis", Sage, London.
- Mitchell W. J. T. (2005), "Imperial Landscape, Landscape and Power.", edited by W.J.T. Mitchell, Chicago (University of Chicago)
- Morata F., Sandoval S. I. (2012), "European Energy Policy: An Environmental Approach", Edward Elgar PubEdward Elgar Pub
- Morrison, M. (2010), "History of SMART objectives. Rapid Business Improvement". Retrieved from http:// rapidbi.com/management/history-of-smart-objectives/.
- Mostafavi M., Najle C. (2003), "Landscape Urbanism a manual for the machinic landscape", AA press, London, UK.
- Mostafavi M., Doherty G. (2010), "Ecological Urbanism", Harvard University Graduate School of Design, Lars Müller Publishers, Baden, Svizzera.
- Nam T. , Pardo T. A. (2011), "Conceptualizing Smart City with Dimensions of Technology, People, and Institutions", The Proceedings of the 12th Annual International Conference on Digital Government Research. https://inta-aivn.org/images/cc/Urbanism/background%20documents/dgo_2011_smartcity.pdf
- Nava C. (2017). "SUSTAINABILITY TRANSITION - Policy proposals and enabling technologies for an open knowledge society". In Monograph.Research REDS 03. Flowing Knowledge, Trento: LISt Lab, pp. 168-171

- Naveh Z., Lieberman A.S. (1990), "Landscape Ecology. Theory and Application", Student Editing. Sprinter-Verlag, New York, USA.
- Nelson A. C. (2002), "How do we know smart growth when we see it". In Smart Growth Form and Consequences, ed. Terry S. Szold and Armando Carbonell, 82–101. Toronto: Lincoln Institute of Land Policy.
- Oswalt P., Overmeyer K., Misselwitz P. (2013), "Urban Catalyst: The Power Of Temporary Use." Berlin, Dom Publishers.
- Palazzo A. (2002), "Identificare i paesaggi in CLEMENTI" A., Interpretazioni di paesaggio, Meltemi editore, Roma.
- Palazzo E. (2010), "Il paesaggio nel progetto Urbanistico", Tesi di Dottorato, Facoltà di Università di Firenze, Dottorato in progettazione Urbana Territoriale e Urbanistica, Firenze.
- Pinna S. (2014). "La falsa teoria del clima impazzito". Edit by: Felici. San Giuliano Terme PI, Italy.
- Ratti C. & Claudel M., (2017) "La città di domani. Come le reti stanno cambiando il futuro urbano." Edit by: Einaudi, Torino, IT.
- Ricci M., (2012) "Nuovi Paradigmi." Edit by: List, Trento, IT
- Ritter. J. (1994), "Paesaggio, uomo e natura nell'età moderna", a cura di Massimo Venturi Ferriolo, Guerini e associati, Milano.
- Rizzi C. (2014), "Quarto Paesaggio", ListLab, Trento.
- Romani V. (1994), "Il paesaggio. Teoria e pianificazione." Franco Angeli, Milano.
- Roger A. (1997), "Breve trattato sul paesaggio", Gallimerd, Parigi.
- Russ. J. (1997), "L'etica contemporanea", il Mulino, Bologna.
- Scarponi A. (2005), "Andrea Branzi: la ville continue [Interview with Andrea Branzi]". Article in Moniteur Architecture AMC no.150 March 2005 / p.88-94
- Schiller F. (2003), "Del sublime" , Abscondita, Collana: Carte d'artisti, Milano.
- Schwab K. (2016), "The Fourth Industrial Revolution", World Economic Forum, Ginevra.
- Sereno P. (1983), "Il paesaggio", in G. DE LUNA (ed.), Il mondo contemporaneo. Gli strumenti della ricerca - 2 - Questioni di metodo, Firenze.
- Shane D. G. (2004) , "On Landscape", The emergence of Landscape Urbanism, Harvard Design Magazine, n.19.
- Shane D.G. (2005), "Recombinant Urbanism: Conceptual Modeling in Architecture", Urban Design and City Theory, Wiley, Chichester, Great Britain.
- Sijmons D., Hugtenburg J., Feddes F., Van Hoor A. (2014), "Landscape and Energy Designing Transition", Published by NAI010 Publishers, Rotterdam hardcover.
- Sommariva E. (2014), "Cr(eat)ing City. Strategie per la città resiliente", ListLab, Trento.
- Taino D. (2013). "Finanza e clima le paure globali", Corriere della sera, 7 Luglio 2013, Milano.
- Tintori S. (1961), "Tony garnier e l sua idea polirica e architettonica per la città industiale", in Casabella, n° 255
- Tomasi C. & Trombetti F. (1985), "Absorption and Emission by Minor Atmospheric Gases in the Radiation Balance of the Earth". La Rivista del Nuovo Cimento, Vol. 8, Ser. 3, n. 2,pp. 89.

- Turri E. (1998), "Il paesaggio come teatro : dal territorio vissuto al territorio rappresentato", Marsilio, Venezia.
- Turri. E. (2004), "Il paesaggio e il silenzio", Marsilio, Venezia.
- Van Timmeren A., Henriquez L., Reynolds A. (2015), "Ubikquity & the Illuminated City", TUDelft, DEI, DIMI, AMS, Delft.
- Waldheim C. (2006), "The landscape urbanism reader", Architectural Press.
- Waldheim C. (2012), "The Landscape Urbanism Reader". Chronicle Books, California USA. p. 23-33
- Waldheim C. (2016), "Landscap as Urbanism : A General Theory", Princeton University Press, New Jersey, UK.
- White M., Przybylski M. (2010), "On Farming: Bracket 1", Actar, Barcellona, New York
- Wiebenson, D.(1969), "Tony Garnier: The citè Industrialle." In: Planeamiento histórico. New York, G. Braziller p. 127
- Zavatta B. (2005), "Per un'estetica della potenza Emerson e Nietzsche sul grande stile." Isonomia, Istituto di Filosofia Arturo Massolo Università di Urbino.

COPENHAGEN
- ASCCUE project. Available online at: http://www.ac.uk/research/cure/research/asccue/ Accessed 7 Jul 2017.
- Adger W. N., Lorenzoni I., O'Brien K. L., eds., (2009). "Adapting to Climate Change: Thresholds, Values, Governance". Cambridge University Press, Cambridge.
- Davoudi S., Brooks E., Mehmood A., (2013). "Evolutionary Resilience and Strategies for Climate Adaptation". Planning Practice & Research, 28:3, 307- 322.
- Eger J.M. (2009) "Smart Growth, Smart Cities, and the Crisis at the Pump A Worldwide Phenomenon," I-Ways32: 1 pp. 47–53.
- Filpa A., Pellegrini V. (2013) "Le potenzialita della green infrastructure per l'adattamento urbano ai cambiamenti climatici". In Urbanistica INFORMAZIONI 252, Anno XXXXI, pp. 6-8
- Furuto A. (2012). "Climate Adapted Neighborhood / Tredje Natur" 26 Aug 2012. ArchDaily. <http://www.archdaily.com/266077/climate-adapted-neighborhood-tredje-natur/>. Accessed 7 Jul 2017.
- GHB Landskabsarkitekter a/s (2014) "TAASINGE SQUARE - rainwater management". Available online at: https://www.ghb-landskab.dk/en/projects/taasinge-square. Accessed 10 July 2017.
- Klimakvarter (2015) "Copenhagen Climate resilient Neighbourhood". Available online at: http://policytransfer.metropolis.org/system/images/1869/original/klimakvarter_ENG_low.pdf. Accessed 7 Jul 2017.
- Klimakvarter.dk (2015). "TÅSINGE PLADS - En lokal grøn oase, hvor regnvand skaber rammer for leg, ophold og nye møder". Available online at: http://klimakvarter.dk/wpcontent/uploads/2015/06/T%C3%A5singeplads_pixi_2015_DK_WEB.pdf Accessed 9 Jul 2017.
- Lindskog H. (2004) "Smart communities initiatives". Available online at: https://www.researchgate.net/publication/228371789. Accessed 10 Jul 2017.

- Mondkar B. (2014) "Cities By Water: What NYC Can Learn From Copenhagen's Urban Planning". Available online at: http://untappedcities.com/2014/05/06/cities-by-water-what-nyc-can-learn-from-copenhagens-urban-planning/ Accessed 10 Jul 2017.
- Nyborg S., Røpke I. (2013) " Constructing users in the smart grid—insights from the Danish eFlex project". Energy Efficiency November 2013, Volume 6, Issue 4, pp 655–670
- Nava C., (2016) "The Laboratory City. Sustainable recycle and key enabling technologies, Quaderni Recycle Italy n.25, Aracne ed., Roma.
- Strzelecka A. Ulanicki B. (2016) "D3.3. Final report combining conclusions from the case studies and the interviews to be used by WP4". Available online at: http://watersoftware.dmu.ac.uk/uploads/publications/d33.pdf Accessed 9 Jul 2017.
- Technical and Environmental Affairs City of Copenhagen (2015) "Copenhagen Climate Projects Annual Report 2015".TMF Design. Cover by Ursula Bach, City of Copenhagen 2015. Available online at: http://kk.sites.itera.dk/apps/kk_pub2/index.asp?mode=detalje&id=1437. Accessed 7 Jul 2017.

BARCELONA

- Clusa J. ; Marmolejo C. (2004), "Barcelona's next post-Olympic urban challenge: The Universal Forum of Cultures 2004." Owned by Istituto Nazionale di Urbanistica; Published by Planum The Journal of Urbanism ISSN 1723-0993
- European Economic and Social Committee, (2016), "Culture, Cities and Identity in Europe Study 20016." Published by: "Visits and Publications" Bruxelles BELGIQUE www.eesc.europa.eu
- Ferretti M. (2017) "Landstocks: New Operational Landscapes of City and Territory.", Publications by List, Trento.
- Gaiddon B., Kaan H., Munro D. (2009) "Photovoltaics in the Urban Environment. Lessons Learnt from Large-scale Projects". Published by Earthscan, London, Sterling.
- OICE Smart City group, (2017) "Smart City: uno strumento per le Comunità Intelligenti." Arti Grafiche srl, Pomezia (IT)
- Salet W.G.M. ; Gualini E. (Hrsg.) (2007), "Framing Strategic Urban Projects: Learning from Current Experiences in European Urban Regions.", Routledge, London, S. xii-307
- Sitography
- Barcelona solar energy. Trailblazer for mandatory solar water heaters. - Consulted 02/02/ 2017 http://wwf.panda.org
- Case Study #3 | Photovoltaic Canopy - Consulted 02/02/ 2017 http://www.mascontext.com/
- South-East Coastal Park - Consulted 02/02/ 2017 http://www.farshidmoussavi.com/node/98
- Eriksson J. - 22@Barcelona – Quad Helix by the Mediterranean. Consulted 10/02/ 2017 blog.bearing-consulting.com
- ECPA Urban Planning - Case Study: 22@ Barcelona Innovation District. Consulted 10/02/ 2017 smartcitiesdive.com

PARIS

- ACEA, (2014) "Carsharing: Evolution, challenges and opportunities". [e-book] Centre for Transport Studies (1st ed), Imperial College, London. Available at: http://www.acea.be/publications/article/sag-report-22-carsharing-evolution-challenges-and-opportunities . Accessed 10 Jan. 2016.
- Autolib'Métropole (2016). [online] Available at http://www.Autolib'metropole.fr/ . Accessed 10 July 2017.
- Corporate Vehicle Observatory (2017) "City Focus Paris 2017 : The Lead To Smart Mobility". Available online at: https://www.corporate-vehicle-observatory.com/news/city-focus-paris-2017-lead-smart-mobility . Accessed 17 July 2017.
- Dowling R., Kent J. (2015), "Practice and public–private partnerships in sustainable transport governance: The case of carsharing in Sydney, Australia". Transport Policy, 40, 58–64.
- Kemp R., Schot J., Hoogma R. (1998), "Regime shifts to sustainability through processes of niche formation: The approach of strategic niche management." Technology Analysis & Strategic Management, 10(2), 175–198.
- Kohrs R., Link J., Mierau M., Wittwer C. (2012), "Charging strategies for a smart home connected battery electric vehicle". Electric Vehicle Symposium 26 Conference Proceedings. pp. 1–6
- Louvet N. (2014), "How does Autolib'users' mobility behavior evolve overtime?"Available online at: http://6t.fr/wp-content/uploads/2015/01/ENAP-anel_ExecutiveSummary_141217.pdf . Accessed 17 July 2017.
- Marty S. (2010), "Autolib' "Prêt-a rouler sur les rues de Paris". Available online at: https://worldstreets.files.wordpress.com/2010/12/Autolib'-sylvain-marty.pdf . Accessed 17 July 2017.
- OECD/ITF, (2011), "ITF transport outlook 2011". OECD Publishing, Paris.
- OICE Smart City group, (2017), "Smart City: uno strumento per le Comunità Intelligenti" Arti Grafiche srl, Pomezia (IT)
- Osei-Kyei R. Chan A. (2015), "Review of studies on the critical success factors for public–private partnership (PPP) projects from 1990 to 2013". International Journal of Project Management, 33(6), 1335–1346.
- Paris Region Economic Development Agency (2013), "Autolib', Two Years Serving Mobility and Smart Transport". Available online at: http://investparisregion.eu/en/news/business-stories/Autolib'-two-years-serving-mobility-and-smart-transport . Accessed 7 July 2017.
- Smith A., Raven R. (2012), "What is protective space? Reconsidering niches in transitions to sustainability". Research Policy, 41(6), 1025–1036.
- Terrien C., Maniak R., Chen B., Shaheen S., (2016), "Good practices for advancing urban mobility innovation: A case study of one-way carsharing". Available online at: http://dx.doi.org/10.1016/j.rtbm.2016.08.001 Elsevier Ltd, 2210-5395/©.
- Weiller C. (2012), "E-MOBILITY SERVICES. New economic models for transport in the digital economics." Available online at: http://www.nemode.ac.uk/wp-content/uploads/2012/12/Emobility-services-final-double-spread.pdf . Accessed 7 July 2017.
- Zaza O. (2016), "Paris Smart City Strategy". Available online at https://eu-smartcities.eu/commitment/4857 . Accessed 17 July 2017.

BORNHOLM

- Bright Green Island (2013), "Bornholm. Bright Green Island". Published by Business Center Bornholm, Rønne, DK. Printed by: niveau2
- Commissione europea. Direzione generale della Comunicazione Informazioni per i cittadini (2015), "Un'energia sostenibile, sicura e a prezzi contenuti per gli europei". Ufficio delle pubblicazioni dell'Unione uropea. Lussemburgo.
- Copenhagen Cleantech Cluster (2012), "Danish smart cities: Sustainable living in an urban world. An overview of Danish smart city competencies." In J. Mortensen, F. J. Rohde, K. R. Kristiansen, M. Kanstrup-clausen, & M. Lubanski (Eds.). Available at: http://www.dac.dk/media/37489/Danish%20smart%20cities_report.pdf . Accessed 19 April 2016.
- European People's Party Group (2011), "Building Europe 2020 in Partnership. Best Practices from Europe's Regions and Cities". Available at: https://www.yumpu.com/en/document/view/19401959/best-practices-from-regions-and-cities-pdf-sign-in-europa . Accessed 07 May 2016.
- Gantenbein D., Binding C., Jansen B., Mishra A., Sundstro O. (2012), "EcoGrid EU: An Efficient ICT Approach for a Sustainable Power System". Available at: http://www.euecogrid.net/images/Documents/130501_ecogrid_ibm_gantenbein-1.pdf . Accessed 25 April 2016.
- Grande O.S. (2015), "EcoGrid EU: From Implementation to Demonstration". Available at: http://www.eu-ecogrid.net/images/Documents/150917_EcoGrid%20EU%20Implementation%20to%20Demonstration.pdf . Accessed 20 July 2017.
- Maring L., Hooijmeijer F., (2014), "Designing with a systems approach". Available at: http://soilpedia.nl/Bikiwiki%20documenten/Snowman/BALANCE%204P/A4-5%20Nirul%20Ramsikor%20Deltares.pdf . Accessed 06 May 2017.
- Trong D. M., Salamon M., Dogru I. (2016), "Experience with Consumer Communications and Involvement in Smart Grid With Examples from EcoGrid on Bornholm". Available at: http://www.eu-ecogrid.net/images/Frontpage/WP-4_final-english-summary.pdf . Accessed 21 July 2017.
- Tyge K., Rikke L. (2015), "Municipalities as facilitators, regulators and energy consumers: enhancing the dissemination of biogas technology in Denmark." International Journal of Sustainable Energy Planning and Management, Vol. 8, No. 2246-2929. pp. 17.

SITOGRAPHY

- www.arcduecitta.it , consulted 10/01/2017. Chiesa A. M. A. (2013), "Landscape Urbanism: un approccio ecologico al territorio urbano"
- www.bilinguescutari.altervista.org , consulted 25/02/2017. Guarente S., Il significato politico filosofico della piazza nella storia d'Italia
- www.rivistadiscienzesociali.it , consulted 25/02/2017. Lai F. (2001), Antropologia del paesaggio: il landscape come processo culturale,
- www.weforum.org consulted 17/02/2017
- www.fr.sogeti.com, consulted 17/02/2017
- www.samefacts.com, consulted 25/02/2017 . Wimberley J.(2006), Prospect e rifugio,
- White M., (2006) "The Productive Surface", www.placesjournal.org, consulted il 20.04.2015
- www.ortobotanicotrieste.it , consulted 10/01/2017, Krasovec L. (2014), Il Giardino tra Natura ed Artificio.
- www.metropolitanstudies.de , Consulted 22/06/2017
- www.stoss.net , consulted 11 April 2016.
- www.climatedata.info
- www.iea.org
- www.eea.europa.eu
- www.time-management- success.com , Consulted 5/06/ 2017.
- www.liferaces.eu , consulted at: 20 January 2015. Progetto RACES Kit didattico sul cambiamento climatico
- www.hr.wayne.edu , Consulted 5/06/ 2017.
- www.setis.ec.europa.eu , consulted at: 17 June 2016 . What is the SET-Plan?
- www.placesjournal.org , (consulted 09 May 2015) . White M., "The Productive Surface"
- www.smart-cities.eu , Consulted 7/06/ 2017
- www.garzantinilinguistica.it , Consulted 20/06/ 2017
- www.weforum.org , consulted 17/02/2017
- www.accenture.com, consulted 17/02/2017
- www.gridinnovation-on-line.eu
- www.addressfp7.org
- www.grid4eu.eu
- www.greenemotion-project.eu
- www.energymatters.com.au
- www.ucd.ie
- www.cer.ie
- www.publications.arup.com
- www.arup.com
- www.theguardian.com
- www.farshidmoussavi.com
- www.mascontext.com
- www.umweltbundesamt.at

SMART LANDSCAPE

Author
Giulia Garbarini

Editorial Director
Alessandro Franceschini

Published by
LISt Lab
info@listlab.eu
listlab.eu

Art Director & Production
Blacklist Creative, BCN
blacklist-creative.com

Editorial Director of LIStLab
Alessandro Martinelli

ISBN 9788898774524

**Printed and bound
in the European Union**,
July 2018

All rights reserved
© of LISt Lab edition;
© of the author's texts;
© of the author's images;

series **BABEL**

No part of this book may be used or reproduced in any form or manner whatsoever without prior written permission, except in the case of brief quotations embodied in critical articles and reviews. Every reasonable effort has been made to contact the rightful copyright owners after their academic course involvement. We apologise for any inadvertent errors or omissions.

Sales, Marketing & Distribution
distribution@listlab.eu
listlab.eu/en/distribuzione/

For more information concerning Listlab's Scientific Boards please visit the webpage: listlab.eu/en/boards/

LISt Lab is an editorial workshop, based in Europe, that works on contemporary issues. LISt Lab not only publishes, but also researches, proposes, promotes, produces, creates networks.

LISt Lab is a green company committed to respect the environment. Paper, ink, glues and all processings come from short supply chains and aim at limiting pollution. The print run of books and magazines is based on consumption patterns, thus preventing waste of paper and surpluses. LISt Lab aims at the responsibility of the authors and markets, towards the knowledge of a new publishing culture based on resource management.